PRIVATE ENVIRONMENTAL REGIMES IN DEVELOPING COUNTRIES

David,

    Thank you for all
the advice and
    encouragement over the
    years.

          Best wishes,
            Raph

PRIVATE ENVIRONMENTAL REGIMES IN DEVELOPING COUNTRIES

GLOBALLY SOWN, LOCALLY GROWN

Ralph H. Espach

palgrave
macmillan

First published in 2009 by PALGRAVE MACMILLAN® in the
United States - a division of St. Martin's Press LLC, 175 Fifth Avenue,
New York, NY 10010.

Where this book is distributed in the UK, Europe and the rest of the world,
this is by Palgrave Macmillan, a division of Macmillan Publishers Limited,
registered in England, company number 785998, of Houndmills,
Basingstoke, Hampshire RG21 6XS.

Palgrave Macmillan is the global academic imprint of the above companies
and has companies and representatives throughout the world.

Palgrave® and Macmillan® are registered trademarks in the United States,
the United Kingdom, Europe and other countries.

ISBN: 978–0–230–61635–6

Library of Congress Cataloging-in-Publication Data

Espach, Ralph H.
    Private environmental regimes in developing countries : globally sown,
locally grown / by Ralph H. Espach.
        p. cm.
    Includes bibliographical references and index.
    ISBN-13: 978–0–230–61635–6 (alk. paper)
    ISBN-10: 0–230–61635–6 (alk. paper)
    1. Environmental policy—Developing countries.   2. Environmental
policy—International cooperation.   3. Environmental management—
Developing countries.   4. Corporations—Environmental aspects—
Developing countries.   I. Title.

GE190.D44E77 2009
363.7009172'4—dc22

                                                          2008052695

A catalogue record of the book is available from the British Library.

Design by Integra Software Services

First edition: August 2009

10 9 8 7 6 5 4 3 2 1

Printed in the United States of America.

CONTENTS

Acknowledgments

Countless individuals contributed to the project in ways both large and small. A couple of bus drivers and an amiable waiter at a favorite Bonairense café come to mind. But the following friends must be singled out for their extraordinary importance.

Several of these are professors at the University of California, Berkeley. Vinod Aggarwal is a friend as well as a mentor. In addition to the education I received from Vinod, his candid recommendations about research and writing–and how one manages to finish research and writing–helped me stay on track when the rails were slick. One of the joys of delving into theories of regulatory policy has been to work with Robert Kagan. Bob is a gifted teacher, and never fails to sugarcoat his critiques and penetrating questions with kindness and encouragement. David Vogel, by engaging me in his ongoing research on corporate social responsibility, demonstrated by example how superior minds tackle complex social and political issues. Finally, as good a guide as she is through the complexities of global environmental politics, Kate O'Neill has been just as good a guide in life, and a friend.

I cannot thank enough my former colleagues and friends at Berkeley. Because of them I will always remember graduate school (erroneously) as fun. They are several, but a few made extraordinary contributions not only to this project, but to my life. Michael Nelson and I seem fated to share oddly construed conference panels and rock-bottom hotel rooms, and to revise each other's chapters and articles on a seasonal basis. I should be so lucky. The benefits I have received from Ed Fogarty's excellent advice on the conceptual framing of my argument are matched only by the rewards of his constant friendship and acerbic wit. If only his hearts game were as keen as his understanding of international politics. Most importantly, it is impossible to imagine what my graduate school experience and this study would be like without Diana Kapiszewski. From Washington D.C. to Buenos Aires to Brasilia, at each intellectual (and logistical) juncture Diana provided sound advice, encouragement, and laughs. I can never thank her, nor any of these friends, enough.

Several individuals showed extraordinary generosity in helping me to overcome the challenges of field research in a foreign country. In Argentina, Pablo Yapura at the *Fundación Vida Silvestre*, Claudia Peirano at the *Asociación Forestal Argentina*; Gerardo Alonso Schwarz at the *Fundación Mediterránea*, and Nelson Culler at *Cuidado Responsable del Medio Ambiente* deserve special thanks for their time and helpfulness toward a research project that in most cases made little sense to them. In Brazil, Gil Anderi da Silva and Luiz Kulay at the *Universidade de São Paulo*, Mirtes Suda at ABIQUIM, and Lineu Siqueira Jr., director of IMAFLORA gave assistance and advice, and introduced me to new ideas and arguments about environmental politics and regulation as much as they did to important officials, managers, and experts within the fields of their expertise. I am also grateful to Resources for the Future, which provided support for the research and writing of this study through a Joseph L. Fisher Fellowship.

This book is the result of an exercise in perseverance and self-discipline carried out quietly, for the most part, at our home in West Seattle. It would never have come to fruition if not for the love and support of my family. Walks with our pointer-lab mutt Maggie bore up my spirits even on the darkest and wettest of days. Naturally, I owe everything to my beloved wife Rachel, who never doubted or complained, and who nodded with well-feigned interest at even the ninety-sixth reformulation of my argument. And finally, I must thank my son Samuel, whose rush at life provided the firmest and most joyful of deadlines.

Ralph H. Espach
November 2008

GLOBALLY SOWN, LOCALLY GROWN: AN INTRODUCTION

> Our industry was established initially in the developed world, but it now encompasses developing nations and economies in transition, and our members in these regions work closely with their local companies to aim for high standards, especially in safety, health, and environmental aspects, in everything that we do.
>
> *Chairman of the International Chemical Councils Association, in a 2006 UNEP report on the global chemicals sector*

> For most companies nothing has changed. It's playing one game instead of another. I tell you one thing. If there is a big industrial environmental disaster, a big accident or scandal, not only in Argentina but anywhere, they will look closely at the preventative system in place, all these audits and certifications, and these programs like Responsible Care, and it will be very bad for all of them.
>
> *Manager at a chemical plant in Buenos Aires, Argentina, interviewed October 13, 2004*

Browse the store shelves, check out the ads on television or in a magazine, and surf the Web: products certified or labeled as environmentally friendly, or *green*, are everywhere. Coffees and chocolates are rainforest friendly. Dishwasher soap is biodegradable or river-friendly, with a stamp from a conservation group. Tuna is dolphin safe, and wild salmon is labeled so that customers can avoid farmed salmon that are linked to river and ocean pollution. Wooden furniture and paper towels come from sustainably managed forests or are chlorine free. Tags on high end clothing explain how bamboo and organic cotton are earth friendly.

In grocery and department stores in the United States, Europe, and beyond, tens of thousands of products bear tags, labels, or stamps claiming production via environmentally sustainable methods. Green products get special display at many retail stores. Music concerts, fashion shows, and other events claim to be carbon neutral, offsetting the emissions they create by paying others to emit less. Each year, dozens of new environmental regimes like these are created, and thousands more products get stamped, certified, or labeled.

What is going on? Have the corporations of the world, those who make the food, the chemicals, the paper, and clothes, all become tree-huggers?

Possibly. To some extent business owners, like anyone else, have grown concerned about the environmental impacts of our industrialized, consumption-driven society and want to do something about it. But of course they could make their products more environmentally friendly and use cleaner, less wasteful production methods without any additional label or certification. Obviously, there is something else besides environmental conscience driving this spread of environmental certification and labeling regimes.

The answer is green of a different kind: there is money to be made. Lots of people around the world now live in sufficient comfort to choose goods and services according to criteria beyond cost alone. They have seen the forests and fields at the edges of their cities disappear. They have walked the banks of rivers or the seashores and had to step over heaps of trash. They have seen, or at least heard of, Al Gore's movie. Aware of the links between what they purchase and environmental degradation around the world, millions of consumers have begun to choose products they can feel good about, even if they cost a little more.

When there is demand, businesses tend to respond. But how can corporations—the same entities responsible for much of the pollution we see and smell, and resistant to rules and laws to clean it up—convince us to trust their newfound commitment to being green? Complying with the law does not cut it. After all, most of the pollution and carbon emissions that have gotten us where we are today were produced legally. Politicians and legislators who depend on industry dollars are as much a part of the problem than of the solution. So how can companies and industries credibly demonstrate that, despite their histories, today they are committed to reducing the environmental costs of their operations and products?

This challenge was first recognized in the late 1980s and early 1990s, as the international environmental movement swelled with boycotts, protests, and the United Nations' Earth Summit in Rio de Janeiro. After spending decades (and hundreds of millions of dollars) impugning environmentalists, many major corporations realized that the facts, and public sentiment, were increasingly against them. They needed a new approach. If they did not take the lead in cleaning up their practices, they would be forced to follow by regulators, lawsuits, and boycotts.

To take the lead, they needed to find ways of showing their commitment to environmentally sound operations. Many businesses tried to do so alone, by printing up statements of principles and codes of conduct and hanging them on their walls. But consumers and environmentalists were skeptical. More progressive companies began to seek out partners, ones with green "street cred." If Greenpeace or the Rainforest Alliance were to endorse their commitment, they thought, who could dispute that?

For their part, many environmental nongovernmental organizations (NGOs) had also begun to look for new ways of promoting environmentally responsible behavior. Protests and boycotts were difficult to sustain and, in the long run, inefficient. In the 1980s and 1990s this need on the part of major businesses, met by environmental groups looking to cooperate, spawned dozens of new initiatives in which private actors designed their own regimes for environmental regulation. The independence of these initiatives was critical to their credibility; government agencies played little or no part. Instead, producers, environmental groups, and certification agencies collaborated to create regimes that were strict enough for environmentalists, yet flexible enough so that companies could participate and still be profitable.

Most of these programs began at the national level, particularly in European and North American countries, where coalitions and partnerships were easily formed and the costs and benefits readily apparent. Several of these, however, spread rapidly around the world. Today private environmental regimes operate within hundreds of industries and in countries worldwide. Forestry, fishing, cars, electronics, chemicals, coffee, and mining all feature environmental standards programs at the global level. The oldest and best established, such as the Forest Stewardship Council, the Marine Stewardship Council, and the chemicals industry's Responsible Care initiative, have spread the farthest. One regime alone—the 14000 series of environmental management standards created by the International Organization for Standardization (ISO)—currently operates across a spectrum of industries.

The proliferation of these private regimes is exciting for several reasons. Considering the weakness of government efforts thus far to address our common, global environmental problems, any new form of regulation is cause for enthusiasm. This is especially true for private global regimes that promise to improve environmental practices in developing states, where industry and consumerism is booming, yet where environmental controls are weakly implemented and enforced.

Moreover, these regimes offer a mode of regulation that reconciles increasing wealth and well-being with environmental conservation. We no longer have to accept that the benefits gained through trade—goods and services that are cheaper and more diverse, as well as economic growth in developing nations—must come with environmental degradation. Instead, free trade may in fact tend to improve the environmental practices of industries. This is because it is trade and investment, especially between rich countries and developing countries, that transfer these standards and rules from country to country.

Private environmental regimes are perhaps most exciting because they present new models for global environmental governance beyond government regulation. In their market-based design, and in some cases like the Forest Stewardship Council in their open, participatory governance structure, these programs are profoundly democratic. People around the world dissatisfied with the feckless, slow efforts of governments to address environmental problems can act themselves, by voting with their dollars, to reward good practices. This decentralized, democratized form of regulation—in which companies and consumers participate voluntarily—could offer a potentially powerful, new instrument for global regulation. This possibility is discussed in detail in the first chapter of this book.

In order to achieve their promise, however, these private environmental regimes must be effective. This requirement of effectiveness takes several forms. First, they must effectively curb the environmental harm caused by the companies and industries they cover. Second, they must do so in ways that encourage efficiency, higher quality, and higher returns, instead of just increasing costs, so that businesses are drawn to participate. A voluntary regulatory regime in which no one participates is an ineffective regime.

Third, they must be effective across a range of political, regulatory, and economic settings. A global regime that operates in only a few countries has only limited significance and could be a trade barrier rather than a trade-neutral device. Many of these private environmental regimes have been most successful in a small number of industrialized countries, especially in Europe. For the most part, it is this success that has brought them to the attention of environmental enthusiasts and analysts of global governance. Their real value, however, their real value as instruments for global environmental governance will depend on their effectiveness in the nations that are industrializing most rapidly and facing the most profound environmental issues. These are the nations of the developing world.

This book asks *to what degree, and under what conditions, are global private environmental regimes effective in developing nations?* All the fancy labeling, advertisements, and lofty rhetoric aside, when and how do they force corporate managers to change their practices, or facilitate their efforts to do so? It is one, very simple, thing to count the countries in which a global regime has an office with a phone number. It is another thing altogether to go to an environmental manager at a factory or on a farm in one of those countries and ask her what she knows about the regime and what, if anything, it has meant to her. This book examines the dynamics of these private regimes within two major developing countries—Argentina and Brazil—through the accounts of managers, certifiers, and regime administrators within those countries.

Analysis of these individual experiences allows us to compare the development and effectiveness of two regimes—one in the forestry and wood products industries, the other in chemicals manufacturing—in two different nations. In the past, analysts of private environmental regimes have compared their general properties, or numbers of members, across many countries. Or they have focused on one regime in a handful of countries, almost always industrialized democracies. This book digs deeper by examining the effectiveness or ineffectiveness of these regimes on the ground, based on the attitudes and behaviors of their local administrators and participating firms. The prevailing wisdom holds that similar trade and investment ties, and the similar presence of transnational corporations and environmental NGOs, lead to similar levels of regime implementation. This analysis, in contrast, argues that the attitudes and institutions of local industries and their associations (what I call their *local organizational capacity*) strongly condition the viability of global private regimes at the local level.

The common wisdom regarding effective regime implementation, and my methods for testing it, are the subject of Chapter 2. In this chapter I present a supply- and demand-side framework for the analysis of the causation behind local regime effectiveness. Most previous studies emphasize demand-side drivers of regime implementation, such as price premiums and access to foreign markets. This study suggests that, under conditions in typical developing countries, supply-side factors are more determinative of regime success.

The two global regimes that are the focus of this analysis operate in contrasting industries and feature distinct operations and administrative coalitions. The Forest Stewardship Council (FSC) is an example of multisector, open participatory regimes typically led by environmental and social NGOs. Many view the FSC, with its open and consensus-based form of governance, as a model for future stakeholder-based regulation.

Responsible Care (RC), on the other hand, is an initiative designed and administered by national and international chemicals industry councils. Industry officials tend to believe that it is chemical engineers, not government officials or environmental scientists, who are best prepared to solve the challenges of finding cleaner, safer ways of producing our everyday goods. Responsible Care is a management standards system promoting continual improvement in safety, health, and environmental practices but allowing individual companies and plants to meet those standards in their own ways. Administered globally by the International Council of Chemical Associations, the regime aims to prove to the public and to governments that the industry can effectively regulate itself and thereby keep the major manufacturers ahead of, instead of behind, the wave of environmental regulation.

Chapters 3 and 5 describe in detail each of these two global regimes, their histories and their structures. Of particular interest is their implementation within South America, for the regional context in terms of regulation and environmental threat can also influence the local demand and supply of these regimes.

Argentina and Brazil make an interesting pair for comparison because they share various economic and regulatory characteristics that common wisdom tells us should promote successful, effective private regimes. These two regimes were introduced in these countries at nearly the same time (1991–1992), by the same foreign firms and transnational NGOs.[1] The conventional wisdom about regime diffusion tells us that these neighboring nations, with similar trade profiles, levels of foreign investment, and economic and regulatory policies, should adopt and sustain these foreign regimes to a similar degree. However, both regimes have struggled in Argentina, while they have thrived in Brazil.

What accounts for this variation, which is consistent across distinct industries and regime types? I argue that national-level factors outweigh the diffusion dynamics of international trade, investment, and penetration by foreign firms and advocacy groups. In particular, the legacies of previous national industrial and environmental policies strongly influence the local effectiveness of these global private regimes.

In the Argentine and Brazilian cases, two legacies in particular have affected regime adoption and development. First, contrasting industrial policies from the 1960s through liberalization in the 1990s shape the structures and profiles of national industries and the dominant culture of firms in ways that influence their receptiveness to foreign models. Second, differences in national experiences with environmental crises affect the mindset with which national businesspeople, individually and within peer associations, approach models of collaborative, voluntary environmental action.

The four case studies that constitute the main source of data for the study are presented in Chapters 4 and 6. These cases show that Brazil's business community, with its diverse and more competitive industries, its tradition of more even-footed integration with foreign capital, and its general recognition of the importance of environmental problems, advantages its firms and NGOs in terms of their capacities for building and supporting successful private environmental programs. In Argentina, on the other hand, a history of stunted state-led industrialization, followed by rapid liberalization coupled with economic and political turmoil, has resulted in weaker and less forward-thinking industry leaders. Combined with a public and civil society less concerned with environmental degradation, in Argentina the firms, NGOs, and individuals that champion private environmental programs do so with far less organizational capacity than their Brazilian counterparts. As a result these private programs are more viable and effective for some producers, sellers, and NGOs than for others, and significant differences exist across industries and national borders, even among countries that in many respects are similar.

Put simply, industry structures and dominant business cultures that are the result of decades-old national policies and experiences shape the capacities of local organizations to coordinate effective local regime chapters. In some countries, national industries have organized themselves and are capable of forming productive partnerships with outside groups, including NGOs. In others, national industries are divided, poorly organized, and defensive before the interests of others in their operations. These differences in local organizational capacity determine the effectiveness, at the level of national industries, of global private environmental regimes.

Globalization does promote the spread of norms, ideas, and institutions, but a nation's economic and environmental history, as well as previous policy choices, condition its receptiveness to these external models. Argentina and Brazil are relatively similar in their policies, economic profiles, and recent histories. As this book shows, however, they are dissimilar in specific ways that, as it turns out, are critical to the functioning of these new regulatory models. Nowhere is this more evident than in the expressed attitudes of business managers and national regime chapter administrators as they describe when and why their friends and colleagues have accepted or rejected these regimes as sources for environmental standards of practice.

Persistent gaps in countries' abilities to implement private standards and certification regimes have important implications for their use as instruments of global governance. Teasing out these implications, at the global level, is the task of the concluding Chapter 7.

First, these national-level differences compromise the uniformity of these regimes. If regimes operate differently across different market settings, they fail to be market-neutral regulatory instruments. Companies may choose to participate in some countries and not in others, based on competitive advantage instead of environmental commitment. Worse, companies that perceive that their national regime is inferior to those in competing nations not only will opt out but will seek to redress this disadvantage by pressing their host countries to file a WTO suit against these environmental regimes as barriers to trade. These private regimes could fall victim to some of the same sorts of thorny trade politics they were meant to overcome.

A second troubling implication is that the factors driving these international gaps are not easily remedied. Legacies of previous industrial and environmental policies have shaped the attitudes and behaviors of a generation of business leaders. These ingrained views will not rapidly change. As the case studies will show, even when governments decide that they want to throw their support behind these new private programs, their efforts can backfire.

So the conditions at the national level that hinder the development of effective private regimes are difficult to overcome.

This book offers several suggestions for the administrators of these global regimes. Most of the efforts of certification and labeling organizations have gone thus far toward building membership or participation rates, or opening new product lines. More recently, mature regimes that have established a significant market presence have turned to the task of boosting demand for labeled or certified products by increasing customer awareness. These activities are important; however, these cases and others in South America indicate that it is equally critical for the organizations and businesses that are stakeholders in these regimes to find ways to supply effective regimes.

This supply-side, local organizational capacity at the level of national industries can take various forms. As three additional case studies presented in Chapter 7 demonstrate, environmental organizations and/or business associations can create innovative, even surprising new coalitions to overcome structural deficiencies. This is the case in Ecuador, where an undersized chemicals industry under duress agreed to implement an environmental management program funded and managed by a leading environmental NGO. This is also the case in Bolivia, where the Forest Stewardship Council enjoys almost excessive acceptance, due to a federal law that makes the regime a central element of national forestry regulation. In cases like Brazil's forestry industry, the strategic targeting of specific firms and individuals as regime participants and board members can allay the fears of other business leaders and give the regime local credibility as something beyond the latest eco-friendly trend.

As with other forms of regulatory governance, to be successful, administrators must strike a balance between flexibility and credibility. The compromises struck among businesses and environmentalists at the global level need to be re-struck among those at the local levels. International standards and procedures must be maintained if the regime is to maintain its reputation among environmental groups and the public. Watering down these systems at the local level only compromises their value. Instead, these cases suggest that a tight adherence to global standards and rules is necessary, and not incompatible with effective regimes, as long as local administrators are able to find ways to expand these regimes beyond regulation to include technical, managerial, and strategic forms of support for participating firms.

CHAPTER 1

PRIVATE ENVIRONMENTAL REGIMES AS TOOLS FOR GLOBAL GOVERNANCE

Before, you could do anything. We never even measured how much effluent we put out. It just went right up the chimney. The used barrels and dirty waste went out with the garbage. If an auditor came we just slipped him a *mango* [payment] and that was all.

But now, whenever there's a sound or a smell, the neighbors come knocking. They don't call the regulators, they come knocking. . . . We need something we can show them. Something credible. So we get certified under these international standards. We want to show that we comply with standards that are as tough as anywhere in the world. And we do it seriously. No one wants to find your factory on some Greenpeace list or read in the newspaper that your plant leaks toxic chemicals.

Environmental Manager at a chemicals plant in Buenos Aires Province,
Argentina, October 2004.

Even within its own borders, a government's control over the behavior of its citizens is never complete. Many regions that defy human settlement—the high seas, deserts, and deep jungle—exist without consistent legal enforcement. In these areas regulation tends to be informal, based on habit or tradition, and is maintained voluntarily among those whose success or survival requires mutual accommodation. On the high seas all ships respond to a distress signal. Informal rules, such as first user's rights, abounded on the American western frontier, as did vigilante justice.

Over the last 40 years, the integration of markets, communications technology, and development have shrunk the world's wilderness. At the same time, however, these processes have expanded extraterritorial, unregulated domains in economic and juridical space. For example, financial markets ebb and flow across the world's computers and communications networks. International derivative funds and speculatory bubbles strike markets globally. People, food, waterways, drugs, and other goods surge across national borders bearing uncertain values, costs, and dangers. The rising volume and speed of international exchanges of goods and services mean that today a greater share of economic activity takes place not inside countries but between them, within juridical space that is not national, but *transnational*.

Transnational corporations are the titans of this sprawling activity. These organizations stitch together the industry, investment, and services of multiple nations into a dynamic network of production and distribution. They shift their assets across countries and oceans in response to changes in cost or demand. The resources of these private entities frequently rival or exceed those of the countries in which they invest and operate, and their operations are far more flexible.

People are understandably anxious about whether their governments are capable anymore of protecting their way of living. Have the powers of states been eclipsed by those of transnational firms or supranational organizations? Or did governments merely delegate some of their authority temporarily, carried away by the wealth that may come via liberal economic policies and integration? Is state power, in fact, as firm as ever, as demonstrated by the authority to tax, print and price money, and wage war?

In capitalist democracies the balance between private and public power has always been tenuous, and has tended to fluctuate. Today, as we face the mounting dangers of global environmental change and persistent international gaps in productivity and incomes, these imbalances are perilous. Across several public policy issues and among populations from Indonesia to the United States to Argentina, free-flowing trade and investment have sparked intense anxiety. In an array of issues including climate change, labor rights, food safety standards, disease control, and human rights, public demands for regulation have gone unmet by governments acting alone or in concert.

This gap between the global expansion of free markets and the abilities of states to regulate those markets effectively is particularly troubling because people around the world are increasingly concerned about events or trends at the global level. The same advances in communications and transport that fuel economic integration also inform people and help them to organize for purposes beyond buying and selling. In recent decades, public demands for more safety, more environmental protections, and greater fairness have risen around the globe. Impatient with the efforts of governments and international institutions, transnational organizations have sprung up to respond to those demands in areas from environmental protection to the elimination of land mines, from workers' rights to intellectual property protection for indigenous peoples.

These nongovernmental organizations pose a challenge to businesses. Should corporations ignore these calls for change and await future regulation via governments, in whatever form that might take? Or should they respond with their own initiatives in order to blunt their impact—or even, in some cases, to create a competitive advantage? Many transnational firms, including most of the world's largest and most successful businesses, have lately chosen the latter path. As a result, in many policy areas transnational governance is increasingly provided via collaborative action among private actors.

WHAT ARE GLOBAL PRIVATE ENVIRONMENTAL REGIMES?

The classic concept of regimes encompasses a variety of institutions that influence behaviors, from formal agreements to shared norms and beliefs.[1] These private regimes that are the subject of this book are formal and have identifiable administrative structures. However, because they operate across hundreds of local markets as well as transnational production and supply chains, different parts of these regimes may demonstrate different degrees of formality and administration. Beyond this—that they are formal, and administered via definable channels and organizations— I further narrow the definition of global private regulatory regimes in the following ways.

First, these regimes operate independently of government agencies or laws. In most cases membership requires compliance with all applicable national and local laws. Especially in developing countries, the technical and operational standards mandated by the regime typically exceed normal local practice. By elevating overall industry performance, industry organizations that conduct their own regimes hope to avoid further regulation or further scrutiny from state actors.

Second, these regimes demonstrate an advanced state of institutionalization.[2] Regimes have defined rules, formal membership, and established administrative offices both locally and internationally. These regimes do more than require compliance with externally derived standards. They define their own standards and procedures for monitoring and verifying compliance. These regimes are not universal; they do not create universal standards, like the ISO's 14000 series of environmental management standards or the self-reporting format offered by the United Nations' *Global Reporting Initiative* (GRI). These regimes are specific to the demands and properties of a given industry or commodity market.[3]

The third distinguishing property of these regimes is their global applicability. Global private regimes are designed to be applicable worldwide, following the reach of whatever market or industry or type of practice that they aim to control. The key challenge that faces administrators, therefore, is to create a framework of basic rules, procedures, and operational structures that assures consistency across national chapters, at least in the pursuance of basic regime principles, but in a manner that is flexible before a range of local conditions. This is a complex undertaking. Regimes must be able to accommodate, in their rules and procedures, different national and industry conditions, both market and nonmarket in character. They also must be responsive to the demands of a range of potential member firms with diverse product lines, management styles, and production processes. Without flexibility before local conditions, the geographical reach of these regimes would fall far short of global, reducing their value. Without responsiveness to the needs or concerns of a wide variety of firms, many firms would be unwilling to join.

To accomplish this combination of consistency and flexibility, global private regulatory regimes typically have a multilevel administrative structure. A global administrative body defines the core principles of the program and the basic structure of its operations, and it coordinates and oversees the program globally. The translation of the regime's principles and system into a working regimen of standards, rules, and procedures is the work of national or regional coordinators.

The key concern of global administrators is to maintain consistency and coherence in the regime's operations across countries and regions. The credibility of the global regime rests, in large part, upon its success at maintaining its reputation for consistency and purpose. Another task of global administrators is to collect reports and data from national chapters and disseminate information about the program. The actual administration of these regimes, however, takes place at the national level. National administrators translate regime principles and rules into practice that is suitable to local conditions.

Global private regimes take various forms, including voluntary agreements to comply with common principles, collective industry self-regulation, independent standards and certification systems, and market-driven product labeling regimes. The variety among private regulatory programs reflects the diversity of issue areas they cover and the interests and parties involved in their design and operation. Some are widely encompassing in their membership and broad in their objectives, such as the Global Reporting Initiative of the UN's Environment Programme or the environmental certification system of the International Organization for Standardization (ISO). Most regimes, however, pertain to a single industry, which generally enhances their technical quality and the specificity of their rules, standards, and verification systems.

Some regimes operate by promoting market-based pressure for corporate accountability through the use of labels on goods or services that meet standards of quality or social or environmental responsibility. Others, such as those based on the concept of corporate social responsibility, support a wide range of interpretations and policies. Each program or regime reflects the ideologies and resources of a specific organizing institution or coalition of supporters, as well as the circumstances in place at the time of its founding and the set of participants at which it is aimed. Chart 1.1 lists several prominent standards regimes in different policy areas.

Chart 1.1   Examples of Private Regulatory Programs

	Origin	Initiators	Policy Problem
Framework for Responsible Mining	2005	Environmental groups, retailers, investors, insurers, and technical experts	Unsustainable and unsafe mining practices
Aquaculture Certification Council	2003	Fisheries, distributors, and retailers of seafood	Environmental impacts of aquaculture, or fisheries
Rainforest Alliance Responsible Coffee	2003	Rainforest alliance	Environmental impacts of coffee harvesting
Fair Labor Association	2003	Industry, Clinton administration	Sweat shops
Program for the Endorsement of Forest Certification (PEFC)	2000	European forest owners association	Unsustainable forestry practices
Marine Aquarium Council (MAC)	1998	Environmental groups, aquarium industry, public aquariums, and hobbyist groups	Ecosystem fisheries management and fish handling
Fair Trade Labelling Organizations International	1997*	Array of NGO and consumer groups	Working conditions, rural poverty
Global Reporting Initiative	1997	United Nations Environment Programme	Need for transparent corporate social responsibility reporting
International Federation of Organic Agriculture Movements (IFOAM)	1997**	Food growers	Food production (soil, water, human health)
Social Accountability International (SAI)	1997	Council on Economic Priorities Accreditation Agency (an NGO)	Workers' rights, community involvement, water and waste
International Organization for Standardization (ISO)'s 14000 series of standards	1996	International association of national standards organizations, with industry representatives	Universal standards for environmental management systems

(*Continued*)

Chart 1.1   (Continued)

Marine Stewardship Council	1996	Environmental groups and Unilever	Fisheries depletion
Sustainable Forestry Initiative (SFI)	1994	American Forest & Paper Association	Sustainable forestry practices
Forest Stewardship Council	1993	Environmental groups, socially concerned retailers	Forest destruction
Sustainable Agriculture Network (SAN)	1991	Rainforest Alliance	Social, labor, and environmental practices
Responsible Care	1986	Canadian and U.S. chemical councils	Safety, health, and environmental regulation of chemical industry
Institute for Agricultural and Trade Policy	1986	U.S. and international small farmers' organizations	Sustainable agricultural practices and free trade

* FLO united 15 separate initiatives, the first of which was the 1988 Fair Trade Initiative based in Holland
** Founded in 1972, but gradually evolved to NSMD system. In 1997 established an arms-length body to accredit certification agencies.
Sources: Bernstein and Cashore (2005), and original research.

THE PROPERTIES OF PRIVATE REGULATORY REGIMES

We can categorize global regulatory regimes according to three properties: their nature, scope, and strength.[4] A regime's *nature* refers to the area of regulation, its design, and its objectives. These elements are interrelated and can generally be traced to the interests or capacities of the parties that participated in the creation of the regime.

The first distinguishing element of a regime's nature is whether it is public (that is, administered by a government or on behalf of a government), private, or a public-private hybrid in which both government and civilian actors take part in its administration. These are not necessarily discrete categories. Many regimes can be usefully compared by the extent to which public and private actors participate in their administration, and the roles played by each type of party. The regimes of interest to this study are both essentially private.

A second important distinction is between *market-based* regimes and *nonmarket-based* regimes. Market-based regimes aim to provide market incentives for companies to comply with their standards. These consist generally of certification systems and product or brand labels so that consumers can identify products or companies that meet the program's product or management standards.[5] Consumers and clients, when purchasing or contracting, can consider the regime's claims regarding the environmental impact caused by the company or product.

Nonmarket regimes solicit voluntary participation by companies and offer means by which companies can indicate a commitment to environmental responsibility, as well as a forum for the sharing of information regarding environmental practices, technologies, and related market and legal considerations. Nonmarket regimes promote compliance with defined standards of management or practice through societal or peer pressures, or in some cases through supply chain pressures, or by emphasizing the benefits that come from more efficient management. These regimes often eschew market-based incentives, partly because achieving public credibility in a market setting requires a level of transparency and independent review with which administrators and participating firms are not comfortable.

Regimes also vary in the nature of their standards and compliance systems. *Performance standards* are based on measurable outputs, such as whether a product meets specific technical standards or achieves targets for pollution or emissions reduction. Certification under the Forest Stewardship Council (FSC), for example, requires compliance with a range of specific technical standards, the stringency of which increases over time. In contrast, *management standards* require only the implementation of management processes designed to lead to improvements in outputs, without specifying performance targets. Management standards tend to be more flexible and are more easily applicable across different types and sizes of firms. Examples of typical management standards include internal systems of measurement and performance data collection, quality control procedures, and open channels for internal communication.[6]

The type of standards a regime establishes is an important element of its design and significantly affects the potential costs and benefits of participation. For example, management standards alone offer no credible basis for product labeling. This is because they do not, by themselves, assure any given level of environmental performance. Responsible Care is one such regime that requires compliance with management—not performance—standards and therefore can make no claims regarding the actual environmental qualities of members' products. Regimes based on management standards alone tend to be less market oriented, since they provide little basis for market distinction and frequently meet with skepticism from outside observers. Market perceptions, however, are subjective. Regimes such as the ISO's 14000 family of environmental standards can sometimes bestow market advantages on certified firms.

A regime's *scope* refers to the range of issue areas it regulates. Regimes may be narrow and focus on only one type of practice or management issue, as with, for example, fisheries management or standards for organic foods. Or they may be broad in scope, encompassing not only environmental standards but also a company's labor policies, community relations, emergency preparedness, workplace safety, financial accounting, product stewardship, and so forth. Narrow regimes run the risk of becoming irrelevant, especially in the face of competitor regimes that integrate more management areas into one standard or certification, thus allowing members to achieve wider benefits more efficiently, in a single audit process.

Comprehensive regimes, on the other hand, have greater administrative and monitoring requirements and can also prove more difficult to communicate effectively to a wider public audience. Regimes wide in scope and flexible in their practice areas may be too undefined in their standards and too inclusive in their compliance requirements to gain broad credibility among outside stakeholders. For example, international initiatives that are designed to encourage corporate social responsibility (CSR) across numerous industries, but without concrete technical definitions of CSR or means of measuring results, fit this description.[7]

Both the FSC and Responsible Care are broad in scope. Each requires compliance with standards covering companies' policies regarding environment responsibility, workplace safety, community relations, and employee health. In addition, the Forest Stewardship Council demands transparency in members' financial and legal dealings, requirements that pose difficulties in countries where laws are not well defined or are inconsistent. Both of these regimes are best known, however, for their influence on participants' environmental or environmental management policies.

A regime's *strength* describes the degree to which it controls the behavior of participating firms. Regimes monitor and enforce compliance with their rules or standards through demands for self-reporting, audits (either by internal auditors or independent auditors), and by offering incentives for compliance and disincentives for noncompliance.

A key indicator of a regime's strength is its method of verifying compliance. Compliance is notoriously difficult to measure, particularly within private regimes where members often create their own benchmarks for its achievement. This study utilizes a useful proxy for regime strength, developed elsewhere, which is the independence of its verification system from the subjects of that verification.[8]

This independence is reflected in a scale that ranges from first-party to third-party. First-party regimes are internal to individual firms, such as self-defined codes of conduct or a corporate pledge. In second-party regimes, member firms' performances are verified by other industry actors, with no role for independent auditors. Responsible Care, for example, is a second-party regime, with minimal transparency regarding firms' performances or scores and all public reporting conducted through national or international chemical associations. Third-party regimes include external verification, usually by certification agencies accredited either by regime administrators or by national or international accreditation agencies. The Forest Stewardship Council features third-party verification by accredited auditors.

Moving along the scale from first- to third-party signifies the increase of verification of firms' performances by external actors. Under a first-party verification system the firm retains total control over its information. Under second- and third-party systems, firms release some control over their management and/or performance data and accept greater degrees of transparency. This increases the risk the company runs if it is not in compliance, or if it is uncooperatives or nontransparent in any way, because outside stakeholders will be monitoring the results of outside audits and assessments. As external verification increases and its independence assured, the regime's credibility improves. For this reason, a regime's or a program's strength has important implications for its legitimacy as a regulatory instrument.

PRIVATE REGIMES AS TOOLS FOR GLOBAL GOVERNANCE

This book argues that despite their purported universal applicability, the strength and scope of private regimes differ dramatically across countries, even those with economic, political, and regulatory similarities. A comparison of four cases in Argentina and Brazil demonstrates that these differences derive from historical variations in national industrial and environmental policies. The legacies of these policies endure in the institutions that pertain to local corporations, industries, and advocacy networks, and in the attitudes of the individuals involved. The same regimes that thrive in Brazil due to effective multiparty collaboration, in Argentina fall victim to distrust and disinterest.

Variation in the effectiveness of private environmental regimes in developing countries speaks to two debates regarding their usefulness as instruments for global environmental governance. One regards their relationship with public environmental regulation at the transnational and national levels. What are, and what should be, the roles of these private regulatory regimes *vis-à-vis* the public sector? Are private regimes spreading because they are better suited than governments to regulate transnational markets? Do they provide a balance between economic productivity and environmental conservation superior to, or more effective than, efforts via national and international law and cooperation? Or is their rapid proliferation due to the powers wielded by corporations and advocacy networks with special influence within today's global economy?

Second, the capacity of these private regimes to provide global environmental regulation depends on whether key audiences accept them as credible and legitimate sources of regulation. Three types of audience must do so: (1) the general public, acting as consumers, voters, and investors; (2) firms and industry associations; and (3) the environmental advocacy community. If any one of these rejects a regime, its legitimacy will be in question. For this reason, an explanation of regime effectiveness must take into account the means by which, and conditions under which, these regimes achieve and maintain legitimacy at the local, national, and global levels.

These issues are conceptually and empirically complex. Analysis must address the actors that are involved, their types and degrees of influence, and the environment within which they interact. To do so coherently requires a theoretical map, one that provides a set of basic assumptions to guide us to key questions. The following section applies two theoretical perspectives—an institutional view and a power-based view—to the questions we have posed about private environmental regimes. These two approaches yield useful, although contrasting, explanations for these regimes and the determinants of their success. Together they provide an analytical framework that helps clarify what we know and do not know about these regimes, the factors that affect performance, and their potential utility as instruments for global governance.

AN INSTITUTIONAL EXPLANATION OF PRIVATE REGIMES

One way to assess the nature of a social organization is to focus on how, and to what extent, the organization serves the interests of participating members or groups of members. Such an approach assumes that these members, or *actors*, are essentially self-interested and via membership in the organization pursue identifiable goals or interests. The organizations or *institutions* (to use a broader term, since people can interact under collective understandings and to collective ends without any formal organization) that actors form serve to help them achieve collective goals. An institutional perspective emphasizes the function of regimes, or institutions, as instruments through which actors resolve problems of collective action.

Institutional explanations for the formation of market-governing regimes stress their importance as tools for overcoming market failures. Market-based regimes can facilitate the operations of markets in many ways: by reducing transaction costs, enhancing informational resources, and stabilizing the market and nonmarket environments that actors face.[9] Regimes improve the efficiency of cooperation and exchange and increase actors' mutual confidence and security. New technologies and other changes in the market or nonmarket environment often engender new challenges. The need to respond to these challenges drives institutional adaptation or the emergence of new regimes.

To understand global private environmental regimes, we must identify the actors that participate in them and the interests and goals of these actors that the regimes serve. Private regulatory regimes are made up of private actors. The private actors involved in the formation and administration of global private environmental regulatory programs are firms and the industry organizations in which they participate, and environmental advocacy groups such as NGOs. These regimes serve distinct functions in the interests of these different sets of actors.

Firms and industry groups create and administer private regimes for several reasons. Most importantly, they do so in response to pressures from the public, advocacy groups, and their own employees and managers for reduced environmental impact. If successful at informing outside stakeholders, especially consumers, about their superior environmental performance via these regimes, firms may hope to benefit in terms of improved brand reputation, market share, or profitability. However, firms and industry groups also create and participate in private regimes to reduce the risk of environmental accidents or damages that can bring severe negative publicity and can invite harsh regulation via state authorities and the courts. By filling the gap between pressures from the public and civil society for superior environmental standards of practice, and from government regulation that effectively mandates such practices, firms hope to make regulation via other actors and channels unnecessary.

By creating and administering private regimes, environmental advocacy groups also aim to fill the gap between public demands and government actions. They do so by concentrating these otherwise diffuse pressures and translating them into incentives for firms and industries to implement superior environmental practices. Dissatisfied with the ineffectiveness of state-based regulation and its tendency to serve the interests of businesses as much or more so than those of the people affected by environmental degradation, these NGOs hope to establish new regulatory systems that are less vulnerable to corporate co-optation.

Of particular interest are the regimes established by or coadministered by NGOs that aim to create and channel market-based pressures toward the end of improving industry practices. These regimes, of which the Forest Stewardship Council is a prominent example, aim to drive normative change worldwide by means of free markets. If successful, these regimes provide a means of overcoming a long-standing perception that the goals of the environmental movement are antithetical to economic growth and development.

Central to the institutional view is the importance of maximizing the efficiency through which collective goals are achieved. In the case of private regulation, efficiency is improved by allowing private actors of different types to play specific roles according to their areas of special competence. Entities such as standards organizations and technical councils provide expertise in the writing of rules and standards. Networks among firms, or firms and external observers, offer instruments of enforcement that are more effective at less cost than top-down state regulation. Efficiency may be improved also by these programs' voluntary, collaborative, and responsive character as compared to the antagonistic relationships typical of regulation "from above" by governments. Private regimes ideally allow firms to innovate and be flexible in complying with the standards of the program, and to think of environmental management holistically and across their entire operations, instead of focusing on how to comply with externally defined technical standard or outcome.

Judging by their own literature, the administrators of private environmental regimes view their role and operations in institutional terms. Corporate environmental managers and NGO advocates use this language when describing their merits: regimes help these parties accomplish shared goals, create win-win situations, and reach mutual agreement on standards and instruments. The metaphors *eco-efficiency* and *green-gold synergy*, found in glossy business literature, as well as the core concept of *sustainable development* that underpins the mainstream approach to environmental policymaking, derive from models of rational behavior and institutionalism.[10]

From this perspective, private environmental regulation is politically neutral and voluntary, and amounts to a collective response among diverse actors to common demands for improved environmental practice. This approach is limited, however, in explaining why programs emerge at a particular time or take the diverse forms that they do. Businesses and environmental advocacy groups may share, to some degree, an interest in responding effectively to the demands of stakeholders for improved environmental practices. Their interest in this end, however, arises from different calculations and values. Businesses like private regulation in good part because it assists and shields them in response to outside pressures. NGOs seek to capitalize on this need to create a system that, ultimately, achieves their objectives of changing corporate conduct.

Actors would not cooperate to form and support private programs if they did not perceive common interests and objectives. Cooperation and mutual benefit, however, are seldom found without conflict at some level, and their gains are never truly equal or evenly felt. Explaining how actors came to cooperate to achieve common goals may be an essential part of the narrative of private environmental regimes. Most often, however, it is also necessary to identify the hands that hold the reins of the regime or of its leading members and thereby determine the terms and structure of their cooperation.

POWER-BASED EXPLANATIONS OF PRIVATE REGIMES

Power-based perspectives on private authority stress the power relations that lie behind the formation and evolution of regimes. Regarding global private regimes, a power-based approach identifies the capabilities and interests of actors (e.g., transnational firms, transnational NGOs, local firms, local governments) and compares their relative influence over regime rules and operations to explain regime outcomes.

A power-based approach aims to look beyond the outcomes of a regulatory act or regime, because such myopia can miss important elements of the story.

A power-based approach assumes that beside, or on top of, shared needs or goals there are dynamics that result from the distribution of power among the interested parties, and these dynamics provide equal or superior explanations for the regimes we observe. For example, a company's participation in a private environmental regime may be motivated by its executives' desire to run a cleaner, more responsible firm. However, the company may also be motivated by the desire to make participation mandatory for all members of an industry association, and by so doing raise the costs on competitors that have less access to capital. The purpose of the approach is to gain important knowledge about these institutions by comparing the relative degrees and types of influence of various actors involved in their administration, the results of that influence, and the distributional effects of program operations, rather than focusing only on their results in terms of environmental practice.

Power-based explanations for the contemporary rise of private regulation differ ontologically and in their units of analysis. A structural approach emphasizes the effects of persistent asymmetries in global markets, in particular balances of power among nations within the capitalist system. In this view, private regulatory regimes reflect the strategic interests of firms and governments of the dominant states in Europe and the United States. For example, one explanation for the proliferation of standards and patent protection regimes is that, faced with declining competitiveness, several major U.S. industries have created these regimes in order to limit market access and raise competitors' costs.[11]

Research in this vein portrays private regulation as yet another instrument for domination by capital-rich core states over peripheral economies. This explains the unidirectional spread of private standards and regimes from the wealthier markets of the North to the developing markets of the South. Developing nations and their producers have often lacked the resources or wherewithal to participate equally in negotiations over the standards and rules of these programs, and receive them as *faits accomplis* despite their impacts on local industries.[12] At a macrosocial level, less-developed countries could be described as forced to negotiate the terms of global environmental policymaking within a conceptual and normative framework defined by, and in service to, the interests of the leading economic and political powers.

Other power-based approaches eschew larger conceptual frameworks of global contention or structural effects, and focus instead on describing the strategies and actions of particular actors, especially transnational firms. Neglected for decades, the interests, strategies, and influence of transnational firms and other nonstate actors have recently come into their own as an analytical theme. With their privileged position as the drivers of contemporary globalization, transnational firms have played important roles in establishing and delimiting the international environmental agenda.

Actor-centered accounts of private regulation are not limited to studies of business strategy. Institutions designed and administered by transnational advocacy groups and networks have significantly affected regulation in areas such as labor rights, human rights, and sustainable environmental practices.[13] The potential exists for advocacy networks and community organizations, made up of common citizens with overlapping local and global interests, to become increasingly active and effective as global actors.[14] Traditionally, business actors and advocacy groups are portrayed as entirely dissimilar and often antagonistic actors. International environmental NGOs such as Greenpeace and Friends of the Earth have effectively targeted transnational corporations and national governments as subjects of public outrage. Historically these groups have more leverage within industrialized nations of the North, where environmental issues are a salient political issue, than in the South, where demands such as jobs creation and crime fighting often overshadow other public concerns.

Decades ago, businesses tended to lobby governments but otherwise remain outside of politics, and activist groups tended to organize protests or boycotts. Increasingly, however, the largest corporations and NGOs engage in the same tactics: coalition building, marketing, lobbying, issue framing, forum shopping, and idea entrepreneurship.[15] Global NGOs tend to their brand image and value, and many major corporations borrow the images and language of grassroots activism to enhance their public reputations. Many environmental NGOs emphasize their pragmatic, solution-oriented approaches to building partnerships and improving global efficiency, while the Web sites for Archer Daniels Midland, Royal Dutch Shell Oil, and General Electric are awash with butterflies, lush forests, and mountain streams. This is a tactical convergence, not an ideological one, a distinction that can be blurred when we focus only on the institutions themselves. A firm that engages in environmental management and pollution reduction still essentially aims to produce and sell a good; an NGO that advocates environmental cleanup and conservation aims essentially at regulation. These are not naturally compatible operations. This conflict between fundamental objectives and their effects on regime design, operations, and evolution appears more readily beneath a power-based analytical lens.

IS PRIVATE REGULATION A CHALLENGE OR COMPLEMENT TO PUBLIC REGULATION?

The transfer of public authority into private hands can be described as the capture of traditional regulatory control. However, the degree to which such a shift involves the loss of state authority or power depends entirely on how these shifts are achieved and how much the new institution shares the same objectives and outcomes as the old. The goals and standards of private environmental regimes, for example, are in concert with formal environmental laws and principles. Most regimes demand full compliance with local laws as a condition of membership or certification. They are designed to complement existing laws and to preclude the need for lawmakers to consider new ones.

Polemical analysts of global governance often portray private environmental regimes as instruments allowing private actors to seize public control. These descriptions overlook both the complexity of internationally competitive markets—in which companies of different types, sizes, and places of origin are constantly attacking monopolies within their industries—and the interdependencies that exist between the private and public sectors. Contrary to abstract models that characterize governments and markets as discrete, exclusionary spheres of control, in reality these two modes of governance overlap and reinforce one another. Private environmental regimes may be formally independent of governments; however, there are many ways by which governments can affect their inception, legitimacy, and operations at the local level. They can make participation mandatory, they can provide tax or regulatory incentives to participate, or they can less formally endorse regime participation. Recent studies have documented the significant effects of government policies or positions on regime outcomes in Europe, North America, and Australia.[16] The same is true internationally as well as nationally, as Miles Kahler and David Lake (2003, 413) describe:

> [There is] . . . an increase in international standards and industry self-regulation in response to globalization. . . . What is less clear, however, is whether such private actions are substitutes or complements for actions by national governments. . . . Although the number of national standards and regulations may appear to decline, the coercive power of the state remains an important actor in the background, one which can be invoked if private initiatives at the international level fail in the eyes of powerful political actors.

From an institutional view, private actors are unlikely to replace public authorities because each group has competencies that are necessary, in their combined application, for markets and development to proceed for the benefit of both.[17] Industry actors have expertise over the technical aspects of production and the management of inputs and outputs that governments cannot match, while governments have the authority to rule that comes from public sanction. Efficiency is enhanced when governments endorse a system in which industry actors can have flexibility in complying with standards and rules that are both achievable and in service to the public good. In addition, NGOs and advocacy networks have their own core competencies as sources of moral authority, expertise, and information. These activists can draw upon diverse sources of expertise, resources, and public activism, and can often spread information more credibly and effectively than can governments or industry groups. They can also provide monitoring at far less cost and more credibly, due to their relative degree of independence from political or market incentives.

Governments, however, remain unique in their powers to provide enforcement in cases of noncompliance. They also have special legitimacy as the source of goal setting in the public's interest, though the credibility and effectiveness of governments in this regard differ across nations and regions. It may be feasible that, under certain circumstances in an area or country with a very poor record of governance, citizens as well as firms may turn to private regulatory regimes—especially those open to their input and participation—as a legitimate source of standards and rules.[18] However, the enforcement powers of private regimes are limited to shaming and expulsion from the club of regime members.

Private regulatory regimes do not seek essentially to replace state regulatory authority, but to overlay formal legal frameworks and provide an additional motivation for standards of performance beyond compliance.[19] Global private environmental programs build upon state regulation by making legal compliance a minimum standard and encouraging integration among ideas of corporate responsibility, superior technical and managerial practices, and legal norms. Regardless of the theoretical labels *public* or *private*, in fact most private regulatory instruments operate as hybrid forms of private-public collaboration.[20]

The diversity among existing private regulatory regimes and their mechanisms for rule making and enforcement complicate any simple categorization. The relations between a given regime and the authority of the state(s) within which it operates demands further empirical research. As we have seen, theoretical explanations of the place of these regimes within global governance, based on either institutionalist rationale or some definition of power relations, are suggestive but not informative about regimes operating today within industries and countries around the world. It is our task to apply these lenses to actual cases, as we shall do in the following chapters.

HOW DO PRIVATE ENVIRONMENTAL REGIMES ACHIEVE PUBLIC LEGITIMACY?

If global private regulatory programs are to endure as instruments of global environmental governance, they must be accepted as legitimate sources of transnational regulatory authority. Legitimate rule making and enforcement is characteristic of governments that, in the modern world, maintain a monopoly on the legitimate use of armed force. Like any government, however, the legitimacy of a national authority rests ultimately on its capacity to provide the public with security, stability, and other services. In a general sense, no other actor possesses the revenue stream—via the powers of taxation—or other capacities comparable to those of modern, bureaucratic states. However, in certain functional areas including environmental protections, national governments have not responded to public demands to the satisfaction of important interest groups. Not only environmental activists, but concerned citizens, businesses, and many special interest groups including hunters and indigenous peoples are dissatisfied with the laws and enforcement in place. In many countries, particularly in the developing world, most national governments fail to implement and enforce the environmental protections that are on the books. Consistent failures of this sort can undermine the government's legitimacy as a tool for public action, compelling these groups to seek alternative governing institutions.

On the other hand, alternative forms of government face the challenge of gaining and maintaining the legitimacy that is commonly granted to national governments. Actors such as industry associations, firms, certification services companies, and transnational environmental NGOs have resources, expertise, and vested interests in creating more effective governing institutions. To do so, however, they must demonstrate to their various members and a broader public audience their ability to act effectively, fairly, and in the broader public interest. Whereas rule making and enforcement by states is broadly accepted, private regimes must convince the community of stakeholders that they offer a credible alternative.

Legitimacy is not seized, it is granted. Legitimate regulation is accepted as fair and necessary by a majority of the target audience of the regulation and, secondarily, by the wider public. Legitimacy is not granted by a single audience but by multiple groups, both concentrated and diffuse.

A private environmental regime must be legitimate in the eyes of potential members and participants—that is, businesses or producers—in order for them to choose to comply with its rules. For these actors *internal* to the regime, it must be neither overly burdensome nor too costly, and must not constrain (more so than the status quo) their capacities for profitable enterprise.

In addition to achieving *internal* legitimacy, private regimes must also convince external stakeholders, such as environmental activists, consumers, investors, other firms affected by environmental rules, and the broader public, that they can offer balanced, effective regulation. Without acceptance from this broader group of stakeholders, a private environmental regime would be merely an agreement among like-minded companies and their partners, without public recognition.

Once this *external* legitimacy is established, a private regime can offer its members a variety of reputational, material, and regulatory benefits.

Environmental activists, potential business partners, and the public view members of a legitimate environmental regime as friends, not enemies, of the public interest. For consumers, private environmental regimes can serve as filters for identifying opportunities when they can "vote" with their money to support products and firms that align with their proenvironmental sentiments. For retailers, downstream clients, and investors, membership in a regime can signal a low-risk, responsibly managed business that would make a sound potential partner. Finally, if a state regulator views regime membership as a credible commitment to good environmental practice, it may focus its attentions elsewhere.

Internal and external legitimacy are mutually supportive. For example, if environmental activists do not accept the regime, environmentally conscious customers are unlikely to perceive it as legitimate. If consumers do not grant the regime legitimacy, then they will not give labeled or certified products or firms any preference, sharply limiting the benefits to firms for participating. If external parties do not recognize or accept a private environmental regime, businesses have little reason to consider membership as a strategy for profitable, sensible environmental policy.

To achieve legitimacy, the administrators of private environmental regimes often engage in various "legitimization tactics." These tactics, which involve refining the regime's design, rules, processes, and/or transparency, can be aimed at both internal and external parties.[21] In order to enhance the regime's credibility, administrators may recruit new partners, such as an environmental NGO, as advisors or overseers. They may devise new rules or mechanisms to enhance their credibility from within. All of these policies and tactics involve the essential balance that regime administrators must maintain between internal and external legitimacy. Private regimes must satisfy their current and potential member firms, who demand flexibility and returns on their investments, and simultaneously satisfy outside parties, particularly environmental activists, who demand accountability and demonstrated effectiveness.

The overarching objective of environmental groups inside private regimes is the same as those outside those regimes: to reduce the environmental effects of industry. Therefore the internal audience of principal interest is made up of firms and industry-based associations, whose long-term objective is to make money. Whether a regime has internal legitimacy depends on the calculations firms make when deciding whether or not to participate, and the depth of their commitment. Essentially a firm must determine if participation would bring profits or savings that exceed the costs of compliance with regime rules and standards.[22] But this is unduly simple. Decisions about participation in a private regime, and the depth of participation, often are also the result of attitudes and personal commitments on the part of executives.[23] Discussions of internal legitimacy, therefore, must encompass not only the microeconomics of environmental management practices but issues of leadership and corporate culture as well. As the following chapters will show, such issues are crucial for explaining differences in the approaches taken by firms and industries in Argentina and Brazil.

Firms and the managers who direct them belong to a community. These communities can be generalized along the lines of the industry or industry sector to which the firm belongs. Thus, the market and nonmarket factors that affect national industries as a whole are another source of influence over private environmental regimes. A large body of research has examined when and why national industries tend to embrace, en masse, private environmental regimes. National industries tend to accept private standards regimes more readily when their export partners do likewise, and when their national regulatory styles are more cooperative and compliance oriented, rather than litigious and top-down.[24] While national trade patterns and regulatory cultures are beyond the power of private regime administrators to affect, their legitimization tactics often respond to particular opportunities or shortcomings derived from national conditions. As we will see, the success of regime administrators at identifying and exploiting the idiosyncrasies of their local market and especially nonmarket environments is a large part of why some private regimes outperform others.

The challenge of establishing external legitimacy is more complicated because external stakeholders are numerous and diverse. Different firms have different external audiences, as do different industries. These audiences include individuals such as neighbors or local community leaders, organizations including local NGOs, community groups, and other business groups, as well as large entities such as firms, governments, and transnational advocacy networks. Common strategies to improve external legitimacy include increasing a regime's transparency, inviting representatives of outside groups to participate as advisors, and involving third-party, purportedly independent groups in the regime's evaluations and compliance verification processes.

Ultimately, external legitimacy and the authority that comes with it depend on convincing external stakeholders that the program is more effective than the status quo, which means in most cases state-based environmental regulation. This is often a tough sell.

First, measuring a regime's effectiveness and comparing it against that of other approaches, real or hypothetical, involve several conceptual and practical complications. External audiences, especially highly committed ones, must often choose to accept or endorse a private regime within dynamic market and nonmarket environments in which alternatives also come and go. Audiences assess not only how a regime operates today and to what effect, but how well it might operate in the future, compared against other types of regulation. The stability of a nation's environmental regulatory policies is, therefore, another important factor affecting the calculations of environmental activists, community groups, and businesses. It is possible that firms inside and outside a private environmental regime may come to prefer the regime's rules and standards to those of local law, or of other regimes, if they perceive them to be more reliable, technically substantive, and effective over the long term. In our case studies of the Forest Stewardship Council, South American firms tended to prefer sound, feasible, and technically precise rules and standards to the more vague language of local and national law, even if the FSC's standards were much more rigorous.

In the eyes of the public, and to many environmentalists, the true test of a private regime's legitimacy is its volume, since this is believed to reflect its overall impact on environmental practice. A simple counting of the firms that participate in global regimes, including the ISO 14001 certification, the FSC, or Responsible Care, suggests that even the largest and best-known private regimes cover only a minuscule share of global production. Moreover, participation in these private regimes tends to concentrate in northern Europe and America, and in specific industry sectors, where state regulation is relatively effective and environmental performance is already quite high. At the same time, the numbers indicate that they hardly exist in other regions or countries such as Africa, Southeast Asia, China, and Russia, where environmental regulation is relatively poor.

Again, however, a cursory count of membership or products covered is an inadequate measure of regime effectiveness. First, the concentration of membership in North America and Europe could be interpreted as a positive sign, since these are also the home markets of most of the world's most competitive and largest companies. Most leading global firms, and those that create trends and models of operation, originate in these markets. Many large European and North American firms also have enormous influence as the drivers of transnational production chains. Indeed, one of the central assumptions of the literature on private environmental programs is that these Northern corporations are principal advocates of and channels for their diffusion worldwide. What GE, Nokia, Cargill, Toyota, and Starbucks do today, firms around the world will likely do in the future.

A regime's small size may also misrepresent its influence. Even with few members and small market share, regimes may shape industry discourse and the attitudes of managers. Regimes spread ideas and knowledge about alternative practices. When successful, they show that alternatives may be profitable and efficient as well as morally sound. Environmental management and corporate social responsibility are now ubiquitous topics in business schools and leading business journals, and increasingly workers at all levels are personally interested in possibilities to improve the alignment between private and public gain.

Although global in their design, the private regimes that are the focus of this book ultimately must prove their legitimacy and effectiveness at the local level, among local audiences. It is not enough to be embraced by the World Wildlife Fund or the World Business Council for Sustainable Development if producers and retailers in countries around the world do not choose to participate. One challenge that faces us, when considering the potential of these regimes, is that questions posed at the transnational or global level—are these programs effective as tools for global environmental governance?—must be answered elsewhere, by examining their effects at the local level. The legitimization process by which private environmental programs achieve authority is fundamentally one of inserting the program and shaping its development within existing national and local institutions and norms. Thus the global effectiveness of private environmental regimes depends upon their achieving and maintaining effectiveness at the local and national level. Our next task, then, requires us to turn to the operations of these private environmental regimes within their local contexts, and to identify the factors and conditions that influence their actual effectiveness.

CHAPTER 2

WHEN ARE PRIVATE ENVIRONMENTAL REGIMES EFFECTIVE AND WHY?

Since the mid-1980s, private environmental regimes have spread across global markets and penetrated local industries around the world, from forestry to fish farms, mining to automakers. By pushing their standards along supply chains and using market and peer pressures to promote their use, regime administrators hope to bring about higher standards of practice more efficiently and effectively than governments can through the traditional tools of public regulation. However, in the eyes of critics, these voluntary regimes do little to change actual practices on the ground, or if they do create change it benefits giant transnational firms and increases their competitive advantages in global markets. Indeed, disagreement over whether and when these regimes are effective often spills over into larger debates over the environmental and economic trade-offs inherent in global economic integration.

The contention over the nature and effects of private global regulation cries out for evidence on the actual impacts of these regimes. Do they improve efficiency, change the attitudes of managers, and reduce the environmental impact from production, or do they not? Do their operations benefit larger, transnational firms at the expense of local companies, or do they not? To assess the effectiveness of private regimes as instruments of global governance, we must first examine their operations and effects at the local level.

MEASURING REGIME EFFECTIVENESS

Questions about when and to what extent international regimes affect the behaviors of nations are central to modern studies of international relations. Scholars have examined why states create institutions, the conditions most favorable to their success, and the causes and consequences of particular institutional designs,[1] but they have struggled to evaluate or compare their effects systematically. In the area of trade policy this research is most advanced, data are abundant, and basic concepts of effectiveness are straightforward (i.e., more trade at lower prices). Yet assessing the strength of an institution is complicated and controversial.[2] The challenge is far greater regarding environmental institutions or regimes, because basic concepts and values are contested, long-term objectives are unclear, and data tend to be scarce and inconsistent.[3] Should the effectiveness of an environmental regime be evaluated according to the stringency of its rules, its rates of compliance, records of enforcement, changes in the behaviors of members, or reduced pollution? How can we identify cause-and-effect relationships regarding phenomena—both corporate practice and environmental conditions—that are constantly changing and where events in one area can have global implications?

To begin, we must think broadly. A regulatory regime can be deemed effective if it (1) changes the behaviors of member parties in an intended direction, (2) solves the problem it was designed to solve, (3) does so in an efficient and equitable manner, and (4) alters the norms and common wisdom of people, inside and outside the circle of members, so that they are more aligned with the principles of the regime.[4] This definition is useful conceptually because of its breadth, yet it would be virtually impossible to apply. Measurements of and comparisons between the values of trade-offs among various desired objectives would be essentially subjective, and changes in thought or attitude are extremely difficult to pin down. A better approach would capture the regime's essential properties, define its objectives in terms relative to its own principles and not to larger questions of fairness or balance, and do so in a way that allows empirical measurement and analysis.

EVALUATING EFFECTIVENESS BY RESULTS

Institutions are created to serve specific functions or to solve specific problems. An important criterion of effectiveness, therefore, must be the extent to which they meet their specified objective. An additional, complex issue to consider is the extent to which the institution in question achieves outcomes superior to those possible via alternative institutions, or via the absence of any institution at all. Environmental regimes generally aim to mitigate the effects of human activities on the natural environment.

Therefore, the greater and more sustained the reduction in environmental impact, the more effective is the regime.

This simple formulation, however, is difficult to apply. First, degrees of environmental impact, and the importance of their reduction, are meaningful only in relation to the absorption capacity of the environment in which they occur. Local environments may be fragile or resilient. Different types of human impact vary in their effects, depending on the environment. Unfortunately, there is no standard for evaluating the capacity of a local environment to absorb or adapt to changing levels of human impact. The interrelations between different types of human inputs over time, and different environmental effects according to local conditions, make any such calculation enormously complex. Also, there is often a significant time lag between the reduction of human impact and the environmental improvements that result. Even in cases where environmental change can be measured, the allocation of causal weight to changes in human practice, and tracing these back to the institution in question, are formidable challenges.

Furthermore, any measure that uses changes in environmental impact as an indicator of regime effectiveness requires some degree of counterfactual analysis. Not only must the analyst explain to what degree the regime in question contributed to the observed changes in environmental impact, but these observed changes must also be compared against the changes that would reasonably have occurred under an alternative regime, or under no regime at all. This type of thought experiment is necessary in order to isolate the role that the regime itself played in determining outcomes from the roles of multiple other factors.

Another challenge is the dearth of reliable data on local environmental conditions, particularly in the developing world. Types of environmental data and methods of collection vary across countries and media (e.g., air, water, soil). Many countries lack any consistent, reliable data at all. This contributes to the asymmetry between the voluminous body of research on environmental regimes conducted in countries of the European Union or Organisation for Economic Co-operation and Development (OECD) and the meagerness of such studies in the developing world.

EVALUATING EFFECTIVENESS BY BEHAVIOR

In the light of these challenges to measuring the effectiveness of environmental regimes by their outputs, many researchers decide to focus on the effects of institutions on actors' behaviors, rather than on environmental impact.[5] However, effects on behavior take numerous forms. A focus on one type of behavior (e.g., reduced CO_2 emissions) risks missing others (e.g., energy conservation or reduced waste) that may be equally or more important over time. Also, indicators based on specific practices may be unreliable because prevailing practices are contingent on other conditions, such as technologies and market conditions.

Measuring effectiveness in terms of specific changes in practice or behavior may overlook important information. Regimes may score low on behavioral indicators in the short term (e.g., members' compliance with requirements, technology upgrades, or the adoption of new management practices), yet achieve goals in the long term by laying the conceptual groundwork for future collaboration. Regimes that appear, by most accounts, to be ineffective may have numerous important effects on a broader scale, for example by raising broader awareness of an issue, spreading information, promoting future cooperation, or changing the way people and organizations view a problem.

Again we face a trade-off between using too narrow a measure for effectiveness and missing other important information, or using a measure so broad that it defies practical application. A middle-of-the-road option, and one that has the advantage of conceptual clarity, is to measure effectiveness according to rates of member compliance with specifically stated institutional rules and standards.[6]

However, using member compliance alone as an indicator of regime effectiveness can be misleading. Low rates of compliance, or the continuation of status quo behaviors regardless of institutional mandates, suggest either an institution that is ignored by its members or one whose standards currently exceed their capacities. In the latter case, the regime could in fact prove highly effective if over time its members come to achieve compliance. High rates of compliance are also ambiguous. Broad compliance may reflect weak rules or low standards instead of high commitment levels. This is particularly true in the case of privately operated regimes, where member companies typically have a hand in writing the rules in order to ensure high rates of compliance and call the regime a success.

Levels of compliance, low or high, require further observation and interpretation in order to explain their relevance to a regime's effectiveness. Quantitative measures at the aggregate level, such as percentages of members in full compliance or absolute numbers of firms that have met specific requirements, are ambiguous without a closer look at members' operations. Compliance alone is a goal or an achieved state; it is not an effect. Compliance is meaningful only when considered in the context of what it means in terms of members' practices and alterations in those practices over time.

A LOCAL-LEVEL APPROACH

This study evaluates the effectiveness of a private regime at the local level by examining its effects on member firms and the operations of its administrators. It compares the national chapters of two regimes in two countries, using data regarding the behaviors of individual firms.[7]

This ground-level approach, with a focus on firms' behaviors and incentives, has several advantages over looking only at compliance rates or membership levels. For one, analysis based on data at the firm or program level reduces the uncertainty involved in the counterfactual claim regarding the effects of this regime as compared against other reasonable alternatives. Focusing on changes in a firm's behavior reduces the variables involved and allows more precise estimations—by examining the incentives and thought processes behind corporate decisions. Facilities managers can more easily compare the effects of regime participation on their own practices, against what they might have been under another regime or none at all, than an environmental expert can compare observed changes at the regional level against other possible types and degrees of environmental change.

Another advantage of local-level analysis is that data from individual firms can be more accurately interpreted within the known context of the local setting. Analysis can more easily account for the difficulty, costs, and significance of reported changes to a specific firm or facility, instead of treating each instance of a similar type or degree of behavior modification as equal. We can even recognize and compare the impacts of different behaviors, practices, or technologies within their specific environmental context, to get a sense of not only the relative difficulty of a firm's alterations in its behavior, but the significance of its environmental effect.

Another advantage of this approach is its flexibility. Analysis does not require equal metrics on specific types of practice or pollution technology across all firms in the sample. Instead, different types and sizes of firms can be usefully compared as long as the data are interpreted clearly and in relation to industrial, technological, and environmental context.[8] Also, because the most pertinent data are gathered firsthand, this project does not depend upon the availability or abundance of secondary data, as do many studies of environmental programs. This makes the measure more suitable for comparative work across developing nations, where reliable data on environmental practice are scarce.

TWO INDICATORS OF EFFECTIVENESS

This study measures regime *effectiveness* along two dimensions: (1) the size and diversity of regime membership and (2) the effects of membership on the environmental practices of member firms. The first dimension reflects usage of the regime and the extent (across an industry) of its effects. The second dimension reflects the strength of the regime and, indirectly, its likely effect on environmental impact. Without participation, any influence the regime may have is irrelevant. On the other hand, without discernible influence over environmental practice, even the most populous regime is ineffective.

High levels of participation do not, alone, ensure effectiveness. Low levels of participation in terms of numbers of members do not necessarily signify ineffectiveness, if the few participants are responsible for most of the industry's impact on the environment. Therefore this first component of effectiveness— regime participation—is assessed using three principal indicators: (1) the number of participating firms, (2) the types of firms represented (and not represented) in regime membership, and (3) the number of participating firms as a share of potential membership. A regime in which only a narrow section of the industry participates is less effective, in terms of membership, than one that is flexible enough in its design and operations to benefit different types and sizes of companies.[9]

An effective private regime is not only one in which a significant and a diverse portion of the industry participates, but one in which these producers' participation drives them to comply with the regime's rules and principles. This second indicator—changes in the behaviors of participating firms— captures two elements of *effectiveness*. It directly reflects the strength of the program at ensuring compliance and indirectly indicates reduction in environmental impact.[10]

Evaluating changes in industry practice poses several challenges. One of these is the likelihood of bias from facility managers' self-reports regarding the extent to which participation in these regimes led to any changes in practice, and the significance of those changes relative to the resources and operations of the firm. The potential for positive bias in this reporting was reduced to some extent in my interviews by explaining that the study was focused on the effectiveness of the national regime chapter, not the firm itself, and by assuring the firm and manager of anonymity. Most managers I interviewed were comfortable discussing the strengths and weaknesses of the national regime chapter and seemed to feel that whether or not membership had affected their firm's practices was not deemed to reflect on the performance of the firm or of themselves as managers.

Using changes in the environmental practices of members as an indicator of effectiveness allows direct examination of the cause-and-effect relationship between participation and changes in practice. Managers, auditors, and regime administrators can report, and often prove via records, what changes came about as results of regime guidelines and pressures, and what changes did not. This is a significant improvement over the practice common in quantitative studies that assumes relationships between observed outcomes and the existence of a regime, without examining this linkage at the micro level.

These two dimensions of regime effectiveness must be analyzed separately because of the possibility of their interrelation. Companies choose whether to participate in a regime based partly on their estimation of the modifications in their practices that participation will require. Under extreme conditions

collective organization, regime administration, and effective partnerships. As with demand-side factors, we can usefully divide these between market and nonmarket types, though the majority of supply-side factors derive from nonmarket actors and conditions.

The principal market-based factor that affects the supply of global private environmental regimes is the presence or absence of verification instruments across the production and supply chains, from the beginning of production to the final consumer. Verification of compliance with regime standards is necessary for the translation of consumer demand for environmentally responsible products or production practices into actual pressure on producers. In the case of commodity or primary goods, these systems often consist of verification of compliance at the production stage and nothing else, since the good is sold in primary form. However, goods that involve more complex, multistage manufacture require integrated systems of compatible instruments at the local and international levels. It is simpler to certify raw wood, grains, or coffee beans—each verifiably grown and harvested according to regime rules—than to certify as a unit the practices and materials involved throughout a car that contains thousands of parts. Transnational verification systems involve organization and investment among regime administrators, firms and industries at each stage of production and/or delivery, and certification organizations. Thus far, only a few industries have succeeded, at least in the environmental field; the wood products industry (including the Forest Stewardship Council) is one of them.

Regime administrators must deal with the problems typical of any collective action, including free ridership, and must establish regime rules that assure the regime's credibility without being excessively rigid. As regimes grow, administrators may face efforts by rivals to undermine their credibility or to create alternatives. How programs at the national level respond to such challenges, and the resources they draw upon to establish their legitimacy, often determine their long-term viability.[16] Only a well-resourced, competent organization can successfully steer a national regime chapter through these institutional challenges, year after year (as our cases in Argentina and Brazil demonstrate).[17] Therefore, one of the most important supply-side factors that determine regime effectiveness at the national level is the competence and capacity of the national administrative body.

The administrators of national regime chapters often seek to increase their credibility by creating partnerships with organizations independent of the industry, such as independent auditors, NGOs, international institutions, or scientific certification bodies. The degree to which these potential partners are present locally, and willing to engage with industry actors on such a project, is a major factor that influences regime effectiveness.

Many regimes, such as the FSC, include industry representatives, environmental advocacy groups, community and labor organizations, and other stakeholders in their management structure. These more transparent and accessible participatory designs improve program credibility, but they can also put firms off by presenting them with new competitive and regulatory risks. Regime administrators cannot allow this openness and inclusion to outweigh firms' rights to privacy and the pursuit of profits. As stated before, a voluntary regulatory regime that pleases environmental advocates, unions, or other interest groups, but to which no firm belongs, is an effective regime.

By definition, private regimes operate independently from the state. As we have seen on the demand side, however, governments can affect the supply of private regulatory regimes in many ways. They can formally endorse private regimes, subsidize regime participation either directly or in the form of reduced licensing costs, or collaborate operationally with regime administrators. Governments may also lend legitimacy to a private regime by incorporating its rules and/or certification into their own practices (e.g., in the management of public forests or state-owned industries), or by making regime participation a criterion in their procurement decisions. Governments may also oppose a regime by creating their own voluntary program or supporting a rival regime that is more friendly to the interests of local industry.[18]

EXPLAINING REGIME EFFECTIVENESS IN DEVELOPING COUNTRIES: FOUR RESEARCH PROPOSITIONS

This demand- and supply-side framework includes the ten qualities of an industry and its market and nonmarket environments that we postulate determine the effectiveness of private environmental regimes. Put simply, for a private regulatory regime to be effective, market and/or nonmarket forms of demand for such a regime must exist and these must be met by supply within the national industry. Supply is often contingent upon an industry's—and a community's—organizational capacity and the conditions it faces within its local regulatory and operating environment.

Ten factors, or independent variables, are too many for clear, direct testing of the causal relationships between each one of the ten and the dependent variable, the effectiveness of a national regime chapter. To resolve this problem, these ten can be grouped into four types: those dealing with market conditions, governments, foreign or transnational firms or NGOs, and industry concentration. The importance of these four types of factor is analyzed by using the research questions that guide this study.

THE IMPORTANCE OF MARKET RETURNS

Most research on private regulatory regimes agrees that one key factor determining their effectiveness is the direct or indirect economic returns they generate for participating firms. As discussed above, market benefits can come in the form of higher prices for certified or eco-labeled products, greater market share, improved market entry, reduced production costs, and lower costs as a result of increased efficiency. If companies do not realize such gains or perceive them as likely, either in terms of increased profit or lowered costs, then participation is extremely doubtful.

In developing nations, market demand for eco-friendly goods tends to be low. For producers in these nations, market benefits from investing in environmentally responsible behavior lie almost exclusively in exporting to Northern markets.

To examine the role that these market forces play in determining the effectiveness of the FSC and Responsible Care in Argentina and Brazil, this study asks: *To what degree do the perceived market benefits of participation in a regime influence the effectiveness of global private environmental regimes?*

Market-based incentives take various forms. Chemical and forestry or wood products firms in Argentina and Brazil may perceive market benefits, as do their counterparts in the North, as less important in terms of sales and prices than for enhancing the firm's image to its clients, consumers, employees, and other stakeholders. Also, because developing-country producers generally operate farther back on the production chain, where their relatively low labor and regulatory costs are a source of competitive advantage, they may emphasize cost savings, price, or market share advantages more than their Northern counterparts. Because of this range of possible manifestations of market demand, observation of the variable *market benefits* must be sensitive to nuance and to the differences in emphasis that firms may place on various types of benefit.

THE IMPORTANCE OF GOVERNMENTS

In industrialized countries, governments have had significant influence over the effectiveness of private environmental regulatory regimes. Governments at the national and local or state levels have influenced private regimes through their roles as buyers of certified goods, regime advocates, regulators promoting or prohibiting labels, and lawmakers incorporating regime standards into local law.

Developing countries, however, often feature weak and corrupt political institutions and relatively high economic and political uncertainty. In many developing countries, environmental laws and their enforcement are largely ineffectual and, when they involve a valuable resource, offer lucrative opportunities to whomever has licensing authority. In many nations, problems of unemployment, poverty, corruption, crime, public health, and education demand the government's attention and push environmental concern to the side. Under tight fiscal constraints, the drive for cost-cutting in areas deemed nonessential to economic growth often hinders the efforts of governments to manage their environmental resources.

The second research question asks: *To what extent do governments in developing nations play key roles in the demand and/or supply of global private environmental regimes?*

This question is open ended because governments have been shown not only to support and promote these regimes on occasion but also to ignore, oppose, or undermine them on others. Regulators or environmental planners may advocate for regime participation because the principles and aims of the regime are in line with public policies. Or they may view environmental certification and improved practice as an element of improved competitiveness, and therefore a target for strategic state support. However, in other countries or in the cases of other regimes, the government may oppose regimes because it perceives them as infringing upon its area of authority or expertise, or because—especially in developing nations—it may feel that they place important local industries or producers at a disadvantage. In our four case studies we will observe both support and opposition on the part of national governments.

THE IMPORTANCE OF TRANSNATIONAL ACTORS

The fact that governments in developing countries tend to dedicate fewer state resources to environmental protection reflects the lack of public support for aggressive action in this area. The fact that environmental conservation or protection often ranks, in less developed nations, as a public priority below security, job creation, health care, and poverty alleviation also affects the size and capabilities of developing nations' environmental advocacy communities. Although environmental advocacy networks in developing nations have grown rapidly in the last 20 years, in most cases they still lag far behind those in Western Europe or North America in terms of their resources, organized constituent base, and political influence.

The growth of organized civil society in these developing nations is also hindered by the weakness of many democratic institutions, particularly the courts and legislative bodies. In addition, markets for eco-labeled goods and demand for environmental certifications have yet to grow to significant size in any developing country. For these reasons, studies of global private environmental regulation have focused on transnational NGOs, firms, and/or business associations as the principal advocates of private regulation in developing states.[19]

To explore the assertion that transnational actors are critical for effective private regulation in developing nations, this study asks: *To what degree does local advocacy for private environmental regulatory regimes on the part of transnational firms and nongovernmental organizations influence the effectiveness of those regimes within developing nations?*

THE IMPORTANCE OF INDUSTRY STRUCTURE

Analysis of private regulatory programs within Northern democracies stresses industry concentration as an important facilitating condition for effective collective action in support of private regulatory regimes.[20] High industry concentration lowers the transaction costs of collaboration and monitoring, and improves information and certainty regarding the behaviors of other firms. Also, high concentration strengthens the capacity of leading firms to play hegemonic roles.

To examine the relevance of industry concentration, this study asks: *Does the degree of concentration in a national industry influence the effectiveness of global private environmental regimes, by affecting the likelihood of collective action among regime advocates and members?*

The simplicity of this last proposition reflects the rationalist assumptions that underpin models of collective action, and the view that firms or other groups form these programs primarily in order to solve problems each could not solve on its own. The challenge is to define the conditions under which these actors are best able to establish and maintain cooperation.

Clearly, there are other characteristics of a national industry that may affect the capacity of its members and their partners to coordinate successfully. Regarding coordination on environmental norms, these may include the industry's history with environmental problems and their resolution; its size, history, and relation to other national industries or policies; and the predominant culture or attitudes of its people. As the cases in Argentina and Brazil will demonstrate, it is necessary to take a comprehensive approach to analyzing a national industry in order to be sensitive to this array of factors. It is also helpful to observe and assess these characteristics at the ground level, among the people and associations who actually conduct the industry's business on a day-to-day basis.

THE RESEARCH DESIGN

This focus on four general factors—market demand, government action, transnational actors, and industry concentration—guides the comparison, in the following chapters, of the effectiveness of the national chapters of the Forest Stewardship Council and the chemical industry's Responsible Care initiative in Argentina and Brazil.

The units of analysis in this study are national regime chapters, embedded within the national industries they aim to regulate. These national regime chapters are not treated as discrete entities but as functioning organizations that emerged via collective action under specific national and international conditions. As we will see, the effectiveness of these regimes is intertwined with the characteristics of their local industries and the political, economic, regulatory, and societal conditions in which they have developed.

The study presents observations of and explanations for complex, multi-dimensional relations that involve a range of actors and interests, several of which change over the period of observation (which is the lifetime of these regime chapters, from the early 1990s until the present). Cognitive and atti-tudinal factors play important roles. This type of exploratory, inferential analysis of multidimensional social phenomena must be inclusive in its approach in order to approximate, in its analysis, any accurate version of the truth. A study like this one, which examines several interrelated independent variables, or factors of interest, requires a small-n research design that focuses on a small number of logically comparable cases. A study of dozens or more cases would confuse the many variables more than it would help to untangle them, and would not allow the type of granular and nuanced analysis that is required in an exploratory study.

The chapters that follow compare the observed impacts of factors related to market demand, state actors, transnational organizations, and industry composition on the effectiveness of two global private environmental regimes in two middle-income, developing nations. The study is both cross-regime and cross-national, with a comparative structure designed to highlight key factors of interest. Variations between the two regimes cast light on the relevance of regime characteristics such as administrative structures and coali-tions of support, and show how these features interact with other factors to influence effectiveness.

The selection of Argentina and Brazil as case studies was intended to serve two purposes. First, these countries' similar economic and political struc-tures, and the parallels in their recent trade and foreign investment policies, help control for national-level factors and allow us to focus on variables at the level of these national industries. Though such natural experiments are never pristine, this focus on two neighbors who underwent democratization and neo-liberal reforms at roughly the same time, and who strive for economic growth chiefly via agricultural and industrial exports, helps clear the concep-tual field for a more granular analysis at the industry level.

Second, both countries feature traits that would theoretically make them receptive to the introduction of global private environmental regimes promoted by transnational firms and NGOs. Argentina's and Brazil's democracies allow citizens, civil society groups, and businesses ample rights to pursue their interests in regard to policy, including the import of private forms of regulation. From the mid-1980s to the present both nations have followed mostly neoliberal economic policy models, which should elevate the salience of international market demand signals.[21] Both countries' forestry and chemical manufacturing industries include significant investment by foreign firms, and have for decades, which should open them to the models and innovations of foreign markets.

Argentina and Brazil both represent "most likely" cases in which, if the prevailing political economic theories regarding the spread of private regulation are correct, global private environmental regimes are likely to be effective. Our puzzle is that simple numbers on these national regime chapters show dramatic cross-national variation in their effectiveness, despite similar national conditions. In Brazil both regimes have flourished and become institutionalized, while in Argentina each has faltered. This suggests that the conventional wisdom, drawn from extensive studies of private regimes in industrialized, wealthy democracies, is erroneous, or at best incomplete, when applied in a developing country context.

A NOTE ON DATA COLLECTION

The information upon which the following analysis is based was collected in Argentina (from August through November 2004) and Brazil (from June through September 2005), via the conduct of semistructured interviews and the examination of hundreds of industry and regime reports. The interviews, 52 in all (29 in Argentina, 23 in Brazil) led me away from my bases in Buenos Aires and São Paulo to the forestry-rich provinces of Entre Rios, Corrientes, and Misiones in Argentina, and in Brazil to several towns around the state of São Paulo as well as to Brasilia. A regional conference of the Argentine forestry industry (held in Posadas, in Misiones province), and both national and international conferences in Brazil regarding the Responsible Care program (held in São Paulo), presented good opportunities for interviews, making contacts, and gathering secondary research materials. In both countries my research benefited from the willingness of dozens of corporate managers, industry association staff members, consultants, certification officials, and others who shared with me their data and impressions on these regimes and the impact they were having on the practices of member companies.

I interviewed regime chapter administrators to learn their perspective on the development of these regimes, their effectiveness, and reasons for their successes and failures. I found these individuals, most of whom work for an industry association or an NGO, to be frank and open about what they felt were the regime's strengths and weaknesses. Most of the interviews, however, were with environmental managers or high-level managers at firms of various types and sizes, both members and nonmembers of the regimes. These interviews yielded the greatest information regarding the impact of the regimes on corporate policies and the various reasons companies have for participating.

I also interviewed officials of environmental regulatory agencies at state (province of Buenos Aires and São Paulo state) and national levels, officials at national standards certification organizations, and staff and directors at several NGOs, both local and transnational. Last, I interviewed auditors at several firms that are accredited to certify companies' compliance with regimes' rules and standards. These individuals were extremely familiar with the details of these national regime chapters, their rules and procedures, and the firms that participate in them. In some cases they proved to be my best-placed and most useful sources.

In order to improve my confidence in my judgment, I asked independent experts to check and comment upon my preliminary analysis. These included former industry officials, former regulators, former auditors, consultants, university professors, and environmental lawyers. A list of the titles and types of organizations of each interviewee can be found in the Appendix. In accordance with a commitment I made to these sources, this study keeps their names and their companies' identities anonymous. In addition to this interview data, I also gathered extensive research from secondary sources such as trade journals, newspapers, and the Internet while in the field and at home.

What is remarkable about global efforts toward forestry management is that widespread public concern, organized activism, and an industry in which many leading producers, distributors, and retailers purport to want sustainable forestry have not led to a functioning international regime. After almost 30 years of public outrage and consistent effort, governance over the world's forest resources and the production and trade of forest products remains patchy and weak.

INTERNATIONAL EFFORTS TO CONTROL DEFORESTATION

The first efforts to create an international forestry treaty emerged in the 1980s, when European and North American environmental groups organized boycotts and public action campaigns against retailers that sold wood from endangered tropical forests. These actions caused major retail chains such as Home Depot and B&Q (a chain in the United Kingdom) to reconsider their suppliers in countries such as Indonesia, Malaysia, and Brazil and to realize that they could not be certain of the origins of their tropical wood products. Facing threats of further boycotts and public campaigns, these retailers quickly became major advocates for improved monitoring and enforcement of forestry laws and practices, either at the national or, more likely, at the international level.

For their part, timber and wood products companies could not ignore this potent combination of pressure from international NGOs and new purchasing requirements from major buyers. Many producers enjoined their governments to fight these emerging attempts at a global convention or forestry treaty. Naturally, governments preferred not to leave this regulation to activist NGOs, private certification agencies, or any other private group. A coalition of developing countries promoted an initiative, led by the United Nations Conference on Trade and Development (UNCTAD), that aimed to create a regime to loosely regulate the production and sale of tropical timber without harming the economic interests of wood-producing nations. In 1986, this effort resulted in the International Tropical Timber Agreement (ITTA) and the creation of the International Tropical Timber Organization, charged with overseeing that agreement. Today 59 nations participate in the organization, but in its 20-year history it has yet to make any clear progress in promoting sustainable forestry or trade diversification.

The 1992 the United Nations Conference on Environment and Development, or the Earth Summit, held in Rio de Janeiro, was intended to spark the institutionalization of various environmental principles and agreements that had been signed but not yet effectively implemented. International forestry officials and forestry control advocates hoped the meetings would lead to the creation of a global treaty on forest management and trade in forest products.

Although the summit failed to generate any binding commitments, the principles that were agreed upon in support of conservation and sustainable development provided the conceptual groundwork for meetings and initiatives throughout the 1990s. Foremost among these were the Intergovernmental Panel on Forests established in 1995 by the UN Commission on Sustainable Development, and the intergovernmental Forum on Forests (UNFF), also a product of the United Nations, created in 1997.

The mission of the Forum on Forests is to develop parameters for a universal legal framework covering forests. Unfortunately, as with the efforts that preceded it, the Forum on Forests has made little progress toward that end.

The work of the UNFF remains mired in bureaucratic concerns over meeting dates and agendas, and virtually no decisions regarding technical requirements of sustainable forestry have been made. Partly in response to criticism of the lack of transparency at the UNFF, in 2001 that body formed a Collaborative Partnership on Forests designed to bring in a wider range of stakeholders to promote existing forestry agreements at the regional and international levels. At a meeting in Japan in 2006, the UNCTAD's International Timber and Trade Organization negotiated a new International Tropical Timber Agreement, expected to come into effect in 2008.[7] However, besides listing new educational and networking projects in support of sustainable trade in tropical wood, the organization failed to address its chronic problem of underfunding. ITTO resources dwindled from US$25 million per year in the early 1990s to under US$10 million in 2005.

Though they have fallen far short of their original goals, these efforts sponsored by the United Nations have fostered dialogue among governments and industry groups and have identified areas of agreement. They also have improved the collection of statistics and information on global forests and the production and consumption of forest products. None, however, has made any significant progress at defining universal technical criteria for sustainable forestry, nor has any led to a binding intergovernmental agreement. Impediments to that aim include persistent differences between Northern and Southern nations over basic objectives and priorities, and contention over technology transfers, land tenure and usage rights, the role of international organizations, and the administration of global forestry institutions.

In recent years, efforts at implementing forest management certification and labeling, sponsored by both environmentalist groups and industry associations, have proliferated worldwide and have come to have a far greater impact than intergovernmental initiatives under the auspices of the United Nations.

THE FOREST STEWARDSHIP COUNCIL

The longest-running program of rules, standards, and verification regarding sustainable forestry practices is the Forest Stewardship Council (FSC). The FSC is an international network of environmental and social NGOs, producers and retailers of forest products, and certification agencies that formulates and promotes standards for environmentally, socially, and economically sustainable forest management. Established in Toronto in 1993, the FSC aimed to unite numerous existing sustainable forestry labels under a single global label so that conscientious buyers could purchase certified wood with confidence. While the FSC seeks to ensure balance among economic, social, and environmental aspects of sustainability, and participation in standards design is open to all parties (with the exception of government officials, who can only participate as observers), the program is generally thought of as an instrument of environmental and, to a lesser extent, community rights NGOs. The World Wildlife Fund, Friends of the Earth, and Greenpeace, for example, have throughout the years been important sources of support and sponsorship.

The FSC system is based on ten global principles, encompassing legal compliance, the rights of workers and communities, reduced environmental impact, and conservation of forests with high biodiversity value. FSC's standards are performance based and specific in their requirements for compliance. The FSC aims to ensure economic as well as environmental and social sustainability, and foresters, wood products manufacturers, and retailers have participated in the FSC since the beginning.

The FSC provides standards and certification of two types, forest management and chain-of-custody control. A forest management certification, awarded by independent auditors trained in the FSC system, assures purchasers that a forest meets established standards for sustainable management. Chain-of-custody certifications, which also come via an independent auditing process, pertain to manufacturing or treatment facilities. These certify that each facility along the production and distribution chain has control over the sources of its wood and maintains its FSC-certified wood separate from wood from other sources. The end result of a series of chain-of-custody certifications is a product label that assures consumers that all of the wood (or a given portion of the wood) used in a final product (e.g., a piece of furniture, a sheet of plywood, a sheaf of paper) originated from sustainably managed forests.

Other social and environmental standards programs use similar certification and labeling approaches. However, the FSC was among the first to define its own principles and to create a process by which different local standards can be established under the umbrella of the global principles. It is this process that has allowed FSC certification to spread worldwide and be flexible to the conditions and requirements of different forests and types of producers, while maintaining the legitimacy of its label. For this reason FSC is considered a pioneering initiative of the "eco-consumerist" approach to voluntary environmental regulation. Since the first set of FSC standards and certifications were available in the mid-1990s, the FSC has certified over 100 million hectares of forest in 79 countries, and several thousands of products carry chain-of-custody certifications.[8]

FSC's international headquarters, based in Berlin, defines and oversees the regime's global principles,[9] accredits certifying agencies, gathers and disseminates information, and monitors program operations around the globe. FSC-International also sets worldwide compliance, auditing, and enforcement procedures. The implementation of the regime's principles, and its general standards, are determined via open votes at annual global conferences, at which environmental NGOs, community groups, and industries have equal representation.

For certified members, audits conducted by an independent, accredited forest certification specialist are mandatory. Audits are required every six months for chain-of-custody certification and annually for forest management certification during the first four years. Oversight of this system is provided by public monitoring and an open investigative process. Any individual or group can call for the investigation of a local FSC certification, a process that consists of outside reviews, requiring full reports from all major local stakeholders.

While FSC-International defines the program's principles, the writing of actual forestry standards and criteria for compliance falls to local working groups coordinated by national chapters. National chapters are in place in 46 countries worldwide. Standards must be defined individually for distinct types of forest such as forest plantations, tropical rainforest, temperate rainforest, alpine forest, and so on. The process of defining local standards consists of negotiations among representatives of economic actors, communities, workers' organizations, and environmental groups who live or operate in those regions. Consensus is required in order for FSC-International to approve the standards. These processes can take several years, or they can last for several years without reaching a final agreement, as has happened thus far in Argentina.

OTHER FORESTRY CERTIFICATION REGIMES

The Forest Stewardship Council is only one of several forest and forest product certification regimes worldwide. Returning to our discussion of regime properties in Chapter 1, these various programs are similar in nature, but differ in their scope and strength. Regarding their nature, all purport to encourage sustainable forest management or practices using certification and in most cases product labeling as their system of verification and reward. Regarding their scope, however, though all involve some definition of environmental sustainability and require legal compliance, they differ widely in the attention they give to labor practices and producers' relations with or responsibilities toward local communities. Likewise, their respective degrees of strength are largely a function of the degree of participation by independent entities in the standards writing and verification processes. Regimes operated by NGOs tend to be the strongest and to require performance-based compliance, while those operated by industry groups and government bodies tend to be more management oriented and flexible in their compliance verification. For this reason it is useful to assess these regimes according to the groups that, for the most part, support and administer them.

NGO-led regimes, such as the Smartwood label offered through a program created by the Rainforest Alliance, are mutually recognized by the FSC. Because of the openness of the FSC governance system, and its support for the objectives and principles of most environmental and social rights groups, it faces few rivals of this type. Most such systems have been subsumed within the FSC. However, industry associations and government agencies, working independently or together and with or without collaboration with environmental organizations or other public interest groups, have created dozens of alternative forest certification programs.

INDUSTRY-LED PROGRAMS

Industry-led programs are intended as more flexible, industry-friendly alternatives to programs such as the FSC that are administered by environmental and community rights groups. The Sustainable Forestry Initiative, sponsored by the American Forest & Paper Association, is an example of a certification program that is essentially controlled by industry.

Industry-controlled standards and certification programs tend to be based on management standards and to feature self-reporting and minimal compliance requirements. From the perspective of environmentalists and community rights groups, these standards and their verification processes are often so weak that they amount to little more than "greenwashing." Their advocates, however, extol the flexibility of these systems in allowing different types and sizes of producers to meet compliance requirements in the manner best suited to their operations.

Industry groups give several reasons for creating rival programs. Many businesses find the standards and certification process of the FSC excessively rigid and expensive. The FSC's deliberative, multiparty processes for standards definitions and dispute resolution are often criticized for moving too slowly and resulting in outcomes that are unreasonable from the point of view of a manager seeking to operate a profitable business. Meeting FSC standards and covering the costs of certification can be particularly difficult for small, family-run or community-based producers with limited technical and financial resources. Also, the FSC's open, deliberative process for the review and revision of policies and standards, held at the local and the global level, introduce uncertainty regarding future requirements.

As the case studies demonstrate, industry associations in developing states have special reasons for concern. To these producers, the global principles of the FSC prioritize the conservationist preferences of Northern environmental and social rights groups over the developmental needs of the South. For example, the demands by some environmental groups to ban from certification all tree plantations growing exotic, nonnative species would prohibit many plantations in the Southern Hemisphere, most of which grow species of eucalyptus and pine that are not native. In addition, the problem that the FSC faces in attracting the participation of small producers is particularly keen in developing nations where forestry is less concentrated and capital intensive than in the North.

GOVERNMENT-LED PROGRAMS

In many cases federal agencies responsible for national technical standards and certifications, which are prohibited from participating in FSC standards negotiations, have created their own forestry standards and certifications. These programs tend to involve close collaboration with industry associations and representatives, and also in most cases with representatives of local environmental and community groups. The Canadian Standards Association program is a prominent example of a government-led program, one that serves as a model for similar programs in Chile and Argentina.

These programs vary widely in the openness, transparency, and inclusion of their standards-setting processes, and their standards and requirements reflect that diversity. Some, like the French national certification program, are essentially industry-led programs with only token involvement by outside groups. Others, like the Swedish or Canadian programs, require independent verification, standards reviews, and performance-based compliance similar to the FSC. These national programs reflect the constellations of political influence over forestry policy in each nation. North European programs tend to stress labor standards, for example, while Canada's program pays special attention to the rights of indigenous groups.

The most prominent global forest certification system besides the FSC is the European Union - based Programme for the Endorsement of Forest Certification schemes (PEFC). Founded in 1999, the PEFC aims to provide an umbrella system, similar to the FSC, to establish mutual agreement among various national forest certification standards programs. To date, the national certification systems of 25 governments have been approved under the PEFC, with a combined certified forest area of over 200 million hectares.[10] The PEFC also recognizes chain-of-custody certifications, allowing it to offer a product certification logo similar to that of the FSC. With enormous certification coverage in Canada and in the United States, where an affiliation exists with the industry-controlled Sustainable Forestry Initiative, and covering the majority of managed forests in Europe, the PEFC is currently the largest forest certification system in the world.

Many observers, however, question the legitimacy of a certification program that has no fixed standards and no minimal performance requirements. PEFC member schemes range from those of Finland and Canada, which share with the FSC open participatory and some performance standards, to those of France, the United States, and Chile, products of national industry groups with minimal reporting and management standards. Administered by the national certification bodies of EU members, European industry associations, and labor unions, the PEFC is criticized for its failure to include civil society organizations and lack of transparency.

Thus far, the program has failed to win widespread confidence in the market. Though some consumers may be savvy decision makers when it comes to wood certification labels, the great majority are incapable of distinguishing the difference between the PEFC and the FSC labels. Of greater weight is the rejection of major, norm-setting environmental groups such as Greenpeace or Friends of the Earth. For the most part these organizations have rejected the PEFC due to the extreme flexibility in its standards and measures, which allows companies and industry associations to choose whatever management standards or requirements would be easiest for them, and call themselves compliant. Critics of the regime are not limited to NGOs. A parliamentary environmental committee in the UK has recommended to its government not to accept PEFC-certified wood as properly meeting sustainable management criteria, because many national systems recognized by the PEFC fail to include any social criteria for sustainable management (House of Commons Environmental Audit Committee 2005).[11]

FOREST CERTIFICATION IN SOUTH AMERICA

Both the FSC and the PEFC have made significant inroads in South American forest and wood products industries. There are FSC-certified forests in every South American and Caribbean country except Suriname and French Guiana. The total FSC-certified area in the continent exceeds 10 million hectares, an area approximately the size of Guatemala. National FSC initiatives exist in Bolivia, Brazil, Chile, Colombia, Ecuador, and Peru (and in Argentina, up until 2007), programs that differ dramatically in their numbers of participating producers, certifications, and degree of development.

Bolivia, for example, has made FSC certification a legal requirement for logging on public lands. Since by the constitution virtually all that country's forests are publicly owned, the vast majority of them are certified under FSC. In terms of the percentage of national forest land covered, Bolivia leads the region and is among the world's leading FSC countries. In 2006 the Brazilian Ministry of the Environment proposed a similar measure. In Brazil's case, the fact that the program has attracted ample participation is lost in the apparently low percentage of total forest that is certified, which is 1.1 percent. However, when one considers that Brazil's total forest area exceeds that of all of Western Europe combined, this 1.1 percent takes on new meaning.

Like Brazil's, the Chilean FSC chapter is well advanced; however, it faces firm resistance from powerful national industry groups. The story in Uruguay is almost reversed. There, the national chapter is relatively new and underdeveloped. Nevertheless, in the last couple of years the country's major tree plantation companies have certified their forests under FSC, so that in terms of size and percentage of national forests, the Uruguay chapter is among the region's leaders. In other countries, the FSC has struggled to generate wide participation or interest. As Chart 3.1 demonstrates, Brazil's national chapter is large, diverse, and advanced in its operations relative to others in the region, and Argentina's is surprisingly

Chart 3.1   The Forest Stewardship Council in Argentina and Brazil compared to the regional average

	Argentina	Brazil	Average of Other South American countries**
Launch of national initiative	2000 (no national standard approved yet by FSC-International)	1994 (first national standard approved 1997)	
No. of certified forests/producers*	12 forests	70 forests	9 forests
	11 chain-of-custody certifications	206 chain-of-custody certifications	11 chain-of-custody certifications
Total hectares certified*	231,126	6,184,118	414,690
Hectares certified as % of total forest area*	.66%	1.1%	4.8%

* As of April 2008.
** Bolivia, Chile, Colombia, Ecuador, Paraguay, Peru, Uruguay, and Venezuela.
Sources: FSC list of certified forests (*www.fsc.org/en/about/documents/Docs_sent*). Total forest area by country: FAO (2000).

underdeveloped, considering the country's large and relatively sophisticated forestry and forest plantations sector.

The Programme for the Endorsement of Forest Certification schemes recognizes as South American members the national certification systems of Chile (CERTFOR, since 2002) and Brazil (CERFLOR, since 2003). As of March 2008 there were only 36 PEFC-recognized certifications in these countries; 16 certify managed forests and 20 are chain-of-custody certifications. The total area certified is just above 2.64 million hectares, around a quarter of the continent's total FSC-certified forest area. The FSC and the PEFC-recognized national forest certification programs together cover approximately 13 million hectares of forest in South America, the vast majority of which is either in Brazil or Bolivia, where FSC certifications dominate, or in Chile, where most are under PEFC certification.

To summarize, the failure of intergovernmental efforts in the 1980s to regulate the wholesale cutting of timber, particularly of hardwoods in tropical rainforests, led consumers and retailers to agree with environmental NGOs that something more had to be done. Along with certification organizations, this coalition created several international—then global—forestry and wood products certification regimes. As of today, the leading two are the Forest Stewardship Council (FSC), which is independent but has strong ties to global environmental NGOs, and the European Union-based Programme for the Endorsement of Forest Certification (PEFC), which differs in its administration across countries. The United States' timber and wood products industries have created their own certification regime, the Sustainable Forestry Initiative, with the support of the U.S. government, but its range is limited to North America.

Both the Forest Stewardship Council and the PEFC have several national chapters within Latin America, most often in the same countries, and they tend to be compared against one another in terms of their legitimacy, rigor, and cost-effectiveness. Indeed, as the case studies in the following chapter will show, they compete within an informal market, as firms, NGOs, community groups, and governments choose which regime best suits their interests without being too costly. In other words, these global private regimes compete in terms of their effectiveness. Such is the case in Argentina and Brazil, to which we now turn.

CHAPTER 4

THE FOREST STEWARDSHIP COUNCIL IN ARGENTINA AND BRAZIL

Popular accounts of globalization depict a world blanketed by networks of new actors that span the globe, beneath which borders disappear. Transnational companies, financial flows, global advocacy networks, and other entities enter local communities and introduce the values, rules, and norms of liberal, postindustrial, capitalist society.[1] Global private regimes are commonly represented as part of this process, outflanking the authority and control of governments, for better or for worse.

This chapter's focus on the Forest Stewardship Council in Argentina and Brazil casts these global regimes in a different light. In these countries, government agencies, foreign companies, and international NGOs have taken a back seat to advocacy groups, industry organizations, and companies at the local level. These organizations, staffed by people with varying levels of talent, energy, and commitment, implement national chapters of the FSC within local and national contexts of cohesion and contention, using whatever organizational resources they have at hand. The FSC is designed to provide a roughly uniform standard of forest management. In reality, however, each national chapter—like each forest—has its own local social and historical context, its own patch of soil, water, and climatic conditions that it must tolerate and from which it must draw strength if it is to survive and grow.

In Brazil and Argentina, the legacies of previous government actions and inactions weigh heavily on the cultures and attitudes of the forestry community. Brazil, once covered from north to south in rainforest, has had a profound national experience with environmental destruction and international condemnation. The Brazilian government also has maintained, throughout the 1980s and 1990s, various state industrial protection and promotion programs. This record contrasts sharply with Argentina's recent history of indifference to environmental problems, aggressive economic liberalization, and political incoherance.

The opinions and experiences of people in these countries' forestry sectors and regulatory bodies, as recounted in these cases, demonstrate the persistent effects of these political, economic, and regulatory legacies. These legacies create attitudes and institutional cultures that shape the behaviors of local corporate groups, regardless of the positions taken by multinational corporations or NGOs. In effect, these local political and environmental regulatory histories strongly condition the implementation of global private environmental regimes.

THE FORESTRY INDUSTRIES OF ARGENTINA AND BRAZIL

The experiences and concerns of FSC administrators and forestry or wood products companies' officials can be understood only within the context of these national industries and their recent histories.

Forestry, or the growth and harvesting of wood, includes two distinct sectors. One involves the more traditional extraction of wood or other forest products (e.g., nuts, rubber, seeds) from forests that grew without cultivation by humans, or what are commonly termed native forests. The other is the cultivation of wood as a crop, which is done via forest plantations or farms.

Forest plantations in Brazil and Argentina grow mostly pine or eucalyptus trees, both fast growing and sturdy. Once harvested, this wood is consumed largely as a commodity, for fuel or for basic construction purposes, or for processing into pulp, paper, charcoal, fiberboard, plywood, compensated wood products, or any of thousands of other products. Tree farming has grown rapidly over the past two decades through foreign investment and the generation of fast-growing, high-quality strains of eucalyptus and pine, both of which are exotic to tropical and semitropical regions. Climatic conditions throughout eastern and southern Brazil and in northern Argentina are excellent for tree cultivation.

With the advantage of favorable climate, and impressive advances in genetic technologies that have created strains of trees specially equipped to thrive in the area, the tree plantation industries in Chile and Brazil are globally competitive. In both countries, cultivation is enormous in scale, production is extremely capital intensive, and the sector is highly concentrated. A handful of companies, such as the Chilean Arauco and the Brazilian Aracruz Celulose, dominate regional production and are increasingly integrated vertically with high-capacity, state-of-the-art sawmills and pulp and paper processing plants.[2] Over the last decade wood, wood products, pulp, and paper have been among the region's fastest-growing exports.

The rapid expansion and competitiveness of the region's tree plantation industries have brought new pressures to bear on the FSC system. The Forest Stewardship Council was first established in North America and Europe, and it is in Europe where the regime is particularly strong. As Southern wood producers improve their competitiveness and begin to take market share from Northern producers, they perceive the forest certification movement as potentially a nontariff barrier blocking their entry into those markets.[3] Industry officials and certifiers, particularly in Brazil, interpret some of FSC-International's recent policy decisions as influenced by the interests of European tree plantation firms and their workers' unions, and designed to reduce the competitive advantage that Southern producers enjoy.[4] According to this view, the current movement within FSC-International to prohibit the certification of exotic tree plantations is driven by fear of competition from Southern forests as much as it is by environmentalist opponents of tree farming—an example of a "Baptist bootlegger" type of proenvironmental coalition. Such a prohibition, if passed, would remove FSC certification as an option for South American plantations based on the large-scale cultivation of nonnative species of eucalyptus and pine.

The second major forestry sector in the region involves extraction from native forests, mostly rainforests in the northern and eastern portions of Brazil and in far northern Argentina. Legal extraction from native forests yields highly valued tropical wood, rubber, nuts, and other products. Hundreds of species of birds, animals, and trees are removed illegally for sale outside of the region. Tropical hardwoods such as teak and mahogany are highly prized for furniture, but after decades of aggressive logging they are extremely rare except in remote or protected forests. The Brazilian Amazon forest is the world's leading source of tropical wood, much of which is logged and sold illegally. Years of intense public concern over deforestation in the Amazon have generated numerous national and local initiatives and institutions to enforce environmental conservation laws, including a ban in 2001 on the cutting of mahogany. The lack of enforcement, however, and the poor application of these measures in general have rendered them ineffective.

Most native forest area in Argentina and Brazil is publicly owned, but poorly managed and protected. The plight of Brazil's forests is well documented: each year thousands of square miles are lost to the expansion of agricultural, ranching, and logging operations, as well as to fires. High international commodities prices, particularly for soybeans and grains, encourage further clearing of forest in both countries. Argentina's tropical native forests, found in the north of the country, are much smaller and less economically developed than Brazil's. The region lacks a tradition of extractive industries comparable to Brazil's harvest of nuts and seeds and rubber tapping; these forests are dwindling due to the lack of economic incentive for their preservation.

The deregulation and market reforms of the late 1980s and early 1990s had a significant impact on forestry in Brazil and Argentina. Under military rule and successive national policies in support of industrialization and development, in neither country did the forestry industry receive the support and protection given to sectors deemed to be of more strategic importance, which include chemicals production. Still, tree plantations and loggers were subsidized throughout the 1980s. In the 1990s, deregulation and liberalization attracted significant investment. These industries grew more concentrated as large firms, including European, Chilean, and North American transnationals, bought the assets of struggling local companies. Different investment patterns in this period proved to have long-term consequences for these national industries. In Brazil, the expansion and concentration of the forestry and wood products sector was driven by the inflow of domestic as much as foreign capital. In contrast, in Argentina investment was led by large, highly industrialized Chilean companies that faced increasing constraints on their operations at home.

Throughout the 1990s the production of wood and wood products, especially pulp and paper, increased dramatically through the use of new genetic lines and more modern, efficient forestry practices. Many higher-value products emerged, both in wood and in paper. However, as the Southern Cone grew more economically integrated and market pressures intensified, the overvalued Argentine peso squeezed national producers. Corporate managers who increasingly operated on a regional, instead of national, scale oriented their Argentine production away from higher-end manufacturing and toward commodity products, many of which would be exported to Brazil. Investors choosing where to build large, capital-intensive pulp and paper plants and wood processing facilities were attracted to the weak Brazilian real, the country's enormous domestic market, and its diverse and deep industrial sector. A virtuous circle emerged in which Brazil's advantages multiplied as foreign investors joined the ranks of already strong Brazilian forestry and wood products firms. In Argentina the opposite occurred. Forestry remained a relatively low-key, traditional industry, centered on the cutting of native wood or the farming of trees on a smaller scale by private landowners.

The East Asian financial crisis of 1997–1998 engulfed both Argentina and Brazil, as international investors suddenly viewed any developing country with suspicion. In response, the Brazilian government allowed the real's value to slide and was granted emergency loans by the U.S. government and multilateral lenders. In contrast, the Menem government in Argentina flatly committed to a dollar-backed peso, regardless of inflation and changes in the values of the currencies of its major trading partners, particularly Brazil. This pushed up the prices of local assets and goods, hurt exports, and magnified Argentina's already staggering debt. Argentinian wood and wood products producers faced precarious times. Economic and political instability in Buenos Aires undermined the efforts of government agencies to stimulate investment or support the industry. Disaster struck in December 2001, when political and economic panic combined to sink the peso, foil a succession of presidents, and demolish the economy. During the year 2002 Argentina's GDP fell by 13 percent, and the share of its citizens living in poverty rose to almost half.

In the last five years both the Argentine and Brazilian economies have grown at a steady—and in Argentina's case an extremely rapid—pace. Lower currency prices combined with sky-high commodity prices on the global market have fueled a boom in exports. The forestry and wood products manufacturing industries in both nations have continued to grow and to integrate their production chains regionally. Nevertheless, the pattern established in the 1980s and the 1990s endures: Argentina tends to produce commodity wood and pulp, much of which goes to Brazil for refinement into high value-added paper or processed wood products, many of which are resold in Argentina.

FORESTRY REGULATION IN BRAZIL AND ARGENTINA

As in other countries around the world, the issue of environmental management and conservation was not a politically salient issue in Argentina or Brazil until the 1980s. Catastrophes like the toxic leak in Bhopal, India, in 1986, which killed thousands of people, and the explosion at the Chernobyl nuclear-powered plant in the Soviet Union were matched by regional crises including deforestation in the Amazon and deadly toxic emissions in and around the Brazilian industrial city of Cubatão. Deforestation in particular attracted intense international attention and as early as the mid-1970s had become a thorn in the side of the Brazilian government. These events sparked public concern at a time when the surge of democratization had galvanized civic activism across the region.

Before the mid-1980s, environmental protection measures in Brazil and Argentina, such as the formation in Brazil of federal Environmental Councils

(Brazil's CONOMA, in 1981), were more symbolic than real. The 1980s brought political transformation across South America, replacing military rule with democratic governments that were willing to confront a range of issues that previous governments had neglected, including environmental protection. Due in large part to the ongoing crisis in the Amazon rainforest, and the Cubatão disaster, Brazil's new government was at the forefront of the region's environmental movement. Its 1988 constitution enshrined the right of all citizens to a sound environment, and in the same year a national Environmental Secretariat was created with its own enforcement agency. In 1991 Argentina created a similar cabinet-level environmental ministry. The structures, missions, standards, and procedures of these agencies were modeled to a great extent after the U.S. Environmental Protection Agency, and much of the early environmental legislation was lifted, sometimes wholly, from U.S. laws. Over time, however, the institutional development of these federal agencies, and the regulatory systems that grew beneath them, were dramatically different in these two countries.

Today, Brazil's national environmental regime is widely considered to be among the strongest in Latin America.[5] The environmental agencies in Brasilia and particularly in the industrial, wealthy states of the southeast possess an extraordinary degree of institutional capacity, resources, and leverage relative to the rest of the region. President Collar in the early 1990s elevated the independence of the national Environmental Ministry by appointing as its director a prominent environmentalist, and subsequent administrations have continued to support—in their rhetoric if not always in their actions—strong environmental management. In some states, such as São Paulo, the environmental protection agencies are as professional and capable as many of those found in wealthy, industrialized democracies. In many others, these agencies are poorly funded and understaffed, toothless in the face of economic development. Also the strength of Brazil's environmental regulatory system varies from issue to issue. The management and regulation of forest activities, for example, is split between the federal agency IBAMA, which oversees all publicly owned forests including most of the Amazon, and state and local regulatory bodies that regulate agricultural activity.

Brazil's environmental legislation and judiciary are remarkably progressive when compared to those of its neighbors. The creation in 1988 of an independent Office of the Attorney General charged with enforcing environmental regulations, among other things, has proved a powerful channel for civil public actions. More recently, a 1998 law that allots individual criminal accountability for environmental malpractice has reportedly had a sobering effect on corporate officials. Under this law, the Brazilian equivalent of state district attorneys (the Ministerio Público office) have become extremely active in many states in the south and southeast.[6] Nevertheless, in most other regions of the country, little has changed.

Argentina's national environmental regulatory regime is relatively weak. The Environmental Ministry created in the early 1990s under the Menem government was subsumed in 1999 into the Health Secretariat, and its budget, staff, and autonomy were cut. Responsibility to control different types of environmental threats (e.g., groundwater pollution, air emissions, agricultural runoff, industrial pollution) was divided across various agencies within the Health Secretariat and others. As a result of this downsizing and fragmentation, many highly trained Argentines dedicated to environmental protection and public health have left the government.[7]

Regarding forestry, the Argentine government's record of regulation and enforcement is especially poor. The technical application and enforcement of national environmental laws is left up to state governments, which in areas where forestry operations are concentrated tend to favor industry and seek investment and economic development regardless of the externalities they may cause. In 1999 the Argentine government passed a law that encouraged investment in forestry by compensating growers for a percentage of their costs. Like many other stimulus policies of that era, this initiative also fell victim to the political and fiscal maelstrom that struck in 2001–2002. Six years on, the government had yet to pay out compensation packages as promised. As we shall see later, the overall fecklessness of national policy toward the forestry and wood products sector, environmental and otherwise, has created a pervasive sense of distrust that today restricts the government's ability to direct any change.

In both Argentina and Brazil, environmental regulation is complicated by the division of responsibilities across different bureaucratic agencies. In addition, due to their decentralized federal systems, environmental regulation in many areas is left to states and local jurisdictions, creating variations in rules and enforcement. Federal law establishes minimal environmental standards and lays out nationwide plans and objectives, but states in both nations differ in the resources and political will they devote to environmental protection or management. Though strict in letter and lofty in rhetoric, many environmental laws have yet to be translated by regulators into technical regulations. As a result many national environmental laws remain vague and impossible to enforce.

Industrial, urban areas tend to be more carefully regulated than rural zones. The highly industrialized Brazilian states of São Paulo and Rio Grande do Sul, for example, have impressive environmental protection agencies. In many rural areas, however, with more interest in business development than in regulation, and where major local industry leaders have strong political ties, environmental enforcement tends to be weak. Moreover, where regulators lack training, satisfactory salaries, and professionalism, permits and licensing systems offer various opportunities for corruption.

THE CASE STUDIES

Chapter 2 identified four factors considered to be instrumental in promoting effective private environmental regimes in developing countries: market factors, governments, transnational firms and NGOs, and the structure of national industries. Market forces affect mostly the demand side of these regimes' operations: do they or do they not raise profits or value, or lower costs? Industry concentration affects the likelihood of supply. State and transnational actors potentially influence regime effectiveness on both sides. They may act as purchasers, market facilitators, informational sources, or other actors that fuel or stifle demand for FSC-certified goods. On the supply side they also may act as key supporters, sponsors, administrators, or opponents, facilitating or complicating the creation and growth of these regimes at the national or local level.

The two case studies that follow evaluate the effectiveness of national regime chapters along two dimensions. The first is their size, which is measured by participation rates and the diversity of their membership. The second is their strength, which is defined as their impact on participants' behavior. As Chapter 2 discussed, the third dimension by which regimes can be assessed—their scope—is irrelevant in this case because these regimes' scope of action is largely defined at the global level.

The analysis is structured to follow the demand- and supply-side framework that is elaborated in Chapter 2. In short, for global private environmental regimes to succeed, there must be demand for certification from consumers, retailers, or the public, and this demand must be perceived by companies. Such demand for private forms of regulation must be met by effective supply of legitimate certification operations, which depends largely upon the capacities of local institutions.

FSC EFFECTIVENESS

SIZE

Participation
No other characteristic of these two countries' FSC chapters conveys so directly their dramatic contrast than the differences in their size and in the diversity, or lack thereof, of their member companies.

Established in December 2001, Argentina's FSC chapter has not yet completed the writing of local forestry standards.[8] The failure of those efforts, which began in 2002, caused FSC-International to withdraw the chapter's status as an FSC national initiative in 2006. Nevertheless, as Chart 4.1 shows, 12 forests have been certified compliant with FSC standards; 7 are owned by small or medium-sized Argentine companies or private landowners. Of these,

4 produce high value-added wood products, such as moldings and floors and medium- or high-density fiberboard, and one produces teas and *mate*, a similar infusion. Not surprisingly, 8 of the 12 are located in the provinces of Corrientes, Entre Rios, and Misiones in the far northeastern region of the country.

Four of Argentina's certified forests are foreign owned, though only one of those units is managed for production. The Argentine forestry and wood products sector is dominated by three transnational corporations, all of which were originally Chilean and two of which remain so. Of these three, only one has sought FSC certification, and then only after the firm was purchased by Gruponueva, a Swiss corporation owned by a global advocate for social responsibility, CEO Stephan Schmidheiny. These three transnational firms operate large-scale plantations and produce mostly pulp and cut wood,

Chart 4.1   FSC-certified forests in Argentina

Company/owner	Nationality	Province	Forest type	Size (hectares)
Las Marias	ARG	Corrientes	Plantation	13,298
FIPLASTO S.A.	ARG	Buenos Aires	Plantation	2,995
Forestadora Tapebicua S.A.	ARG	Corrientes	Plantation	6,906
Forestal Santa Bárbara S.R.L./The Candlewood Timber Group LLC	US	Salta	Native forest	81,332
LD Manufacturing S.A. LIPSIA	ARG	Misiones	Plantation	3,892
UBS Brinson FIDEICOMISO Financiero forestall	Swiss	Corrientes	Plantation	16,146
Agrupación Bosques Libres Mendocinos	ARG	Mendoza	Plantation	229
ECOBOSQUES, Corporación Ecológica y Bosques Tropicales S.A.	Spain	Corrientes	Native forest	202
Garruchos S.A. – Estrella del Bosque S.A.	ARG	Corrientes	Plantation	22
Fideicomiso Santo Domingo	ARG	Corrientes	Plantation	3,340
Agrupación Bosques Libres Mendocinos	ARG	Mendoza	Plantation	229
Forestal Argentina S.A.	Chile	Entre Rios, Corrientes	Plantation	44,986

Source: Forest Stewardship Council registry, as of May 2008. Available at *www.fsc.org*.

product lines in which there is little demand for certification.[9] One obstacle to their participation is FSC's prohibition of certification for any plantation established after 1994 upon land that previously was native forest, which rules out certification on much of these companies' lands.

The rest of the Argentine forestry industry consists of a handful of medium-sized, relatively capitalized firms and an estimated 2200 local farmers and sawmill operators selling logs or cut wood, mostly for the domestic market (Braier 2004; *República Argentina* 2002).[10] Three collectives of small producers account for the two most recent FSC management certifications.

Since the creation of the Working Group in 1996, FSC-Brazil has had three different sets of standards approved by FSC-International (for forest management in the Amazonian rainforest, for extractive use of forests in the Atlantic Forest, and for the management and harvesting of cashew fruits), and two others await approval. Brazil's certification numbers are among the highest in the world. As of May 2008, 70 Brazilian forests, totaling 6,184,118 hectares, are FSC certified, and 206 Brazilian companies hold chain-of-custody certification. Brazil's participation in FSC resembles more closely that of Canada (40 forests, 23,592,610 hectares) or the United States (103 forests, 9,975,780 hectares) than other Latin American nations.[11]

Because of the immense size of Brazil's total forest area—comparable to the whole of continental Europe—even 6 million hectares, an area larger than the whole of Costa Rica, seem insignificant as a percentage of total forest area (approximately 1.1 percent). Of this total area of certified forest, FSC certifications account for 96 percent. The remaining 4 percent pertain to certifications under CERFLOR standards, an industry-led system similar to Chile's CERTFOR program (and to Argentina's proposed government-sanctioned system).

Brazil's planted forests sector is highly concentrated. As of 2006 approximately 20 companies operate in the sector, but 4 firms stand out in terms of forest assets and production volume.[12] In sharp contrast to the Argentine case, where Chilean and European-owned plantation and pulp and paper firms dominate the sector, most of these firms in Brazil are Brazilian or of joint Brazilian-foreign ownership. As of 2007, approximately 60 percent of all planted forests are certified under FSC, and an estimated additional 10–15 percent are preparing for certification. One company official interviewed in 2005 predicted that, between FSC and CERFLOR certifications, by 2010 as much as 90 percent of managed or planted forests will likely be certified.[13]

Participation Rates Over Time

Ideally, Argentina's FSC chapter could be viewed as being at an early stage of development. If that were true, participation would increase as national standards are established and as early-mover firms demonstrate the feasibility and benefits of certification. However, the apparent trend of reduced interest in FSC certification over time (shown in Diagram 4.1 does not bode well for its future development.

The majority of all existing certifications were obtained early, from 2001 to 2003, without regard for the lack of nationally defined FSC standards. As with any voluntary standards program, many of these early certifiers were firms or private forest managers worthy of the label "true believers,"[14] who were using superior forestry practices before FSC certification became available to them. This was certainly the case with *Las Marias* and *Alto Verde*, for example, two producers for whom environmental stewardship has been a principle of their ownership and an element of their image since their inception.

A critical stage in the evolution of private environmental programs is reached when secondary potential participants, neither committed to nor opposed to investing in superior standards and certification, decide to participate because they are convinced that doing so is worth the costs and effort. Thus far FSC-Argentina has not reached this stage. Instead the Argentine industry seems to be in a wait-and-see period, watching how local FSC administrators respond to recent setbacks (most of which—like the national economic collapse—were beyond their control) and how the market for certified wood develops.

FSC-International's removal of the Argentine chapter as a formal initiative reflects the fact that private certification regimes, at least in their early years,

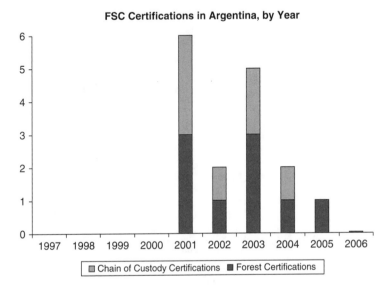

Diagram 4.1   Forest certifications in Argentina, by year

must grow and improve consistently in order to survive. There is no evidence that global or regional market demand for certified wood products decreased between 2002 and 2006. In fact, it grew. The faltering of the Argentine FSC chapter, therefore, is likely the result of problems of a supply-side nature.

In contrast, the data presented in Diagram 4.2 indicates that in Brazil the perceived value of FSC certification has increased over time, not only on the part of forest operators but up the production chain as well. FSC-Brazil's eight-year history of consistently rising certification numbers suggests that far more than early movers or "true believer" firms have chosen to participate.

Forestry plantations tend to be large-scale, capital-intensive operations, obvious to observers and to the law. However, most native forest logging and extraction is small in scale and conducted illegally, but extremely difficult to regulate. An estimated 85 percent of all wood shipped from the Amazon is cut illegally.[15] In contrast to the high percentage of tree plantations that are certified, less than one-half of 1 percent of total native forest area is certified.[16] Still, Brazil's FSC chapter boasts the world's largest area of certified tropical forest in the world. With a tropical rainforest half the size of the continental United States, the Brazilian Amazon forests offer tremendous potential for future certifications.

Brazil is not only the world's largest supplier of tropical wood; it is also the world's largest market. Only around 35 percent of wood logged in the Amazon is sold for export. The rest is sold domestically, more than half

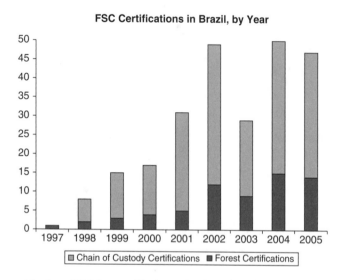

Diagram 4.2   Rate of FSC forest certifications in Brazil (as of March 2006)

within the state of São Paulo. At present, the most significant impediment to further certifications of Amazon rainforest is the lack of clear legal ownership and tenureship rights in the region, which limits the growth of legal operations and reduces incentives for certification. A federal Program for Forest Management, passed into law in 2006, was designed to clarify land rights on public lands and encourage certification in the Amazon. Thus far the effects of the program in terms of certification and the expansion of legal, regulated production are difficult to discern.[17]

STRENGTH

Effects on Certified Companies' Practices

FSC certification in Argentina demands appreciable modifications in management practices. Company managers interviewed estimated that the initial costs of the certification process, including changes in practice or technology needed to reach compliance with FSC standards, ranged from US$50,000 (for smaller operations) to over US$150,000.

The challenges posed by certification varied across firms of different sizes and regions, but most reported that requirements for the legalization, training, and care of the workforce are the most difficult to achieve. FSC's requirement of demonstrable, full legal compliance raises costs significantly in a sector where most hiring and contracting are informal and where many competitors operate illegally.[18] Moreover, laws and requirements differ across federal and local agencies, and often pose unreasonable obstacles to compliance.[19]

FSC's criteria for fair wage, training, benefits, and workers' safety often pose more difficulty for companies than those regarding environmental practices. Plantation operators that emphasize quality control for specific product lines tend to have more stable workforces and receive the greatest returns from investment in worker training. Their higher profits, in general, help support investments in worker satisfaction and productivity. On the other hand, these costs seem to fall heaviest on small producers, which lack resources for such investment and are highly vulnerable to market swings and other forms of uncertainty. FSC's social standards are also difficult for the largest companies, such as one of the large Chilean companies, which outsources its operations to 53 contractors and has a total primary and secondary employee pool of around 4000 people, within a region rife with informal labor and contractors. The scale of these complications is one factor behind the firm's reluctance to certify under FSC.[20]

The institutional mechanisms behind the reported "seriousness" of FSC certification are its international requirements for accreditation,[21] regular audits, and public transparency. The program's credibility is also based on its international status. As one industry official phrased it, "do you think [the FSC] would compromise its international legitimacy . . . for some bribe from some small Argentine businessman?"[22] FSC's credibility in Argentina rests largely on these properties of the system, not on any direct enforcement. FSC-Argentina has not conducted any official review in Argentina nor has it reprimanded any certification agency. With a part-time staff of one, the chapter has nowhere near the resources for proper, regular monitoring and verification. Any enforcement of this type must be conducted via short-term interventions by FSC staff from other countries.

In Brazil, the impact of FSC certification on producers' behavior differs across the two sectors of the forestry and wood products industry. Plantation operators report that FSC certification has had little impact on environmental management or practices. This is due not to lax FSC standards but to the fact that most of Brazil's forestry companies had diligent, high internal standards of management and practice well before FSC certification became available in 1997. Managers report that obtaining FSC certification was relatively easy. Years of scrutiny from environmental groups, as well as a general awareness of the national problem of deforestation, had already pushed industry norms high above those in other Latin American countries.[23]

Among Brazilian plantation managers, the greatest reported impact of certification regarded these firms' relations with local communities. The most frequently remarked difference between the standards of the FSC and the more industry-friendly CERFLOR system is that those of the FSC are more stringent regarding the building of policy consensus with local communities.

For extractive operations in native forests, the impacts tended to be far greater. These operators report significant changes in environmental practice, community relations, and operational accounting. Many extractive operations are run by communities or families and are small in scale. To many of these seed, wood, fiber, and rubber harvesters, modern forms of record keeping, data storage, and practice management are completely new. Obtaining an FSC certification entailed dramatic investment in training, new equipment, planning efforts, and the research of different extraction techniques. Larger Amazonian operations, on the other hand, report relatively little impact from FSC certification. Similar to their counterparts running large tree farms, these large-scale operators tended already to manage their forests with care in order to distinguish themselves from the untold tens of thousands who exploit Amazonian forests illegally and irresponsibly.[24]

As with plantation managers, firms that operate in native forests report that the impact of FSC certification is greatest in the areas of labor management, community relations, and social development. Many managers report that environmental management, which is basically technical or procedural, is simpler to communicate to workers and to implement than are improved social and community management, which require skills and training beyond those typical of forestry technicians. Also, improvements in environmental practice are sometimes made easier by the availability of technical and financial assistance from state agencies, universities, consultants, and certification providers.

FACTORS THAT INFLUENCE THE EFFECTIVENESS OF THE FSC

DEMAND SIDE

> So far there is only limited interest in FSC certification, and this is mostly from the high-end manufacturers. For them, FSC pays off in exports. For companies that sell commodities like logs, unfinished wood, or pulp it does not make sense because there's no market for certified commodities. . . . Some [managers] say that the only reason they survived the crisis [in 2001] was because of certification. But these companies, unfortunately, are not the norm in Argentina[25].

Market Demand for Certification in Argentina

In Argentina, the foremost incentive for firms to seek FSC certification is to gain access to markets and clients in Europe and North America. Firms that hold FSC certification report that the label has proved a key advantage in tapping those markets. There is no domestic demand for certified wood or wood products, nor do firms report any gains from certification within their other main trading partner, Brazil.

Market benefits from the FSC label are reported to be greatest in the refined wood products sector, where sales outside of Mercosur countries (i.e., to Europe, North America, or to Asia) were highest, and for companies that sell eucalyptus wood products, in order to overcome the negative image associated with that tree among European buyers.[26] On the other hand, none of the producers of pulp or untreated wood, destined for domestic or regional markets, has yet sought FSC certification.

Argentine firms report very few, if any, market benefits from FSC certification. Only one company official interviewed claimed that certification had brought demonstrable gains in terms of profit. All others, certified and noncertified, claimed that no such premium exists. One representative of a company with FSC certification expressed disappointment that, after two years, their offering of certified product had failed to have any significant impact on their sales.[27] Still, the company planned to maintain its certification with the hope that this would change in the future. Many experts view certification as a gamble, especially for small companies. It entails significant short-term costs for uncertain longer-term benefits of being a leader and innovator in penetrating markets for higher-quality wood. Many producers are watching and waiting to see if the investment in FSC certification will ever pay off for those already certified.

Another source of market demand for FSC participation comes from corporate clients, principally transnational corporations, which either prefer or mandate that their suppliers and partners have FSC certification. In the case of FSC-Argentina, no such supply-chain pressure is reported to exist. Transnational firms still influence the implementation of the FSC. European and U.S. firms generally support the FSC and other private environmental regimes, and they account for a significant portion of certifications. The two remaining Chilean firms, which still dominate the wood products industry in the northeast, are the FSC's most obdurate opponents.

Likewise, demand from retailers does not yet exist in Argentina. The weakness of downstream demand is evidenced by the country's relatively few (11) FSC chain-of-custody certifications. In 2002, the cancellation of a Home Depot project in Buenos Aires disappointed several FSC-certified firms.[28]

This lack of market demand or price premiums for certified product limits the attractiveness of certification and strengthens the position of firms that oppose the FSC. One Argentine industry consultant explained:

> When [a large Chilean firm] came here, they told everyone here that they're not fools like the people who cave in to pressure and certify. They've gone almost ten years, they said, and despite all the pressure there was no reason at all to certify. There were no market returns. They said they'd wait another ten years to see if it's worth it then. . . . This was the attitude these companies brought to Argentina.[29]

Nonmarket Demand for Certification in Argentina

Pressures from state regulators or NGOs, or threats of litigation, are reported to be absent or insignificant. State regulations and enforcement over forestry practices are extremely weak and easily avoided. Aside from the organization that coordinates the FSC, Argentine environmental NGOs focus more on stopping deforestation, or on urban and public health issues, than on sustainable forest management. Despite constitutional protections for public natural resources, the law does not permit collective injury claims in lawsuits. This complicates any litigation against companies over forest degradation. As yet there have been no cases of successful litigation against forest owners or companies on grounds of environmental damage.

Market Demand for Certification in Brazil

The chief reason that Brazilian managers gave for seeking FSC certification was the expectation of improved market access and price premiums. As in the Argentine FSC case, however, these expected benefits are rarely realized. For firms that export to Europe, certification is reported to be an increasingly common criterion for market entry. Price premiums, however, are rare and differ by market segment. Premiums of 20 to 50 percent have reportedly been paid for tropical native wood carrying the FSC logo. As a result, 70 to 80 percent of Brazil's native wood exports are FSC certified.[30] However, for cut wood, fiberboard, or compensated wood products, the FSC label brings no significant price premium due to inexpensive supplies from within the EU or from Asia. Prices for pine board are determined mostly by swings in global supply and demand, and FSC certification brings no meaningful price premium. Even in these markets, however, certification is an increasingly important criterion for market access, especially in Europe.

Domestic demand for certified wood and wood products is limited. For producers of cut wood, pulp, chips, and most paper products, consumer or client demand for certified goods hardly exists. Some industry experts claim that FSC-certified wood pulp can bring a small premium (of around 1 percent) in certain product lines, but company officials disagree. One producer scoffed at the claim: "A price premium for certified goods? Forget it! Certification adds costs, more than benefits."[31] For many product lines including pine planks, charcoal for steel mills, or wood as fuel for pizza ovens (an enormous source of demand, especially near the voracious city of São Paulo) there is little prospect of consumer or client-based demand for certification. Some firms, however, have found niche export markets—such as selling charcoal briquettes or high-end paper—in which certification has provided a competitive advantage.[32]

The key reason why most Brazilian firms maintain their certification is that, in a national industry with as bad a reputation as Brazilian forestry, responsible firms must do everything they can to indicate their social and environmental responsibility. Since the 1970s, Brazil's national image has been tarred by images of ruthless deforestation and an uncaring government. This legacy places a mighty burden on companies that wish to legitimately sell products from the rainforest. They must differentiate themselves from a sea of nefarious, unethical competition and compete against the constant supply of cheap illegal wood. Many firms ultimately view certification as critical for their public image, or to avoid scrutiny and criticism. One CEO in the western Amazon sought out FSC certification because of pressures from within her family. "I needed something to tell my daughter when she came home from school and asked how I know that my company is not killing the forest."[33]

The operators of Brazil's massive tree farms should have an easier time of it. First of all, most of this cultivation and harvesting takes place in the south and southeast, thousands of miles from the Amazon. Tree plantation operators like to point out that their production reduces demand for wood from native forests. Nevertheless, they also suffer the stigma of Brazil's inability to staunch the loss of its native forests. The entire forestry industry in Brazil has been "under a regulatory and public microscope for years," because of the symbolic importance of the nation's forests.[34] Some plantation officials who harvest eucalyptus and pine wood in the south of Brazil report facing reluctance on the part of foreign clients to buy any Brazilian wood that is not certified. Those retailers want to see the FSC logo in order to protect themselves from any connection to Amazonian rainforest wood.

The rapid and sustained growth in FSC-certified forests and wood-processing operations in Brazil has led to the happy circumstance of a steady domestic supply of certified wood. Now, industry analysts have turned their attention to the problem of a lack of market demand for all the certified wood available for sale. If certified products cannot find any market opportunities different from noncertified products, firms may begin to reconsider the costs of maintaining their certificates. For example, one firm that produces FSC-certified cardboard claims that certification actually hinders, instead of helps, market access. Potential clients that manufacture boxes or other packaging stay away from FSC-certified cardboard because, to maintain the value of the certification, they would need to obtain chain-of-custody certification for their operations. Since there's no market return, it is easier for them to avoid certification altogether and deal only with noncertified product.[35]

Because Brazilians buy more tropical wood products than any other country in the world, FSC administrators and advocates see great promise in the promotion of domestic demand for certified products. Now that there is a significant internal supply of certified wood and wood products, there is growing pressure to increase domestic demand for certified products.

To build a client base, FSC partners have organized national groups of certified wood buyers and sellers. These associations host annual fairs and encourage networking among harvesters, processors, sellers, and consumers. The World Wildlife Fund has formed an association among top-end architects, furniture designers, and builders who use only certified wood. This group, Design and Nature (*Desenho e Natureza*), began holding annual exhibitions featuring certified wood products in 1999, and after 2000 only FSC-certified wood products were permitted for display. The strategy is to target the use of certified wood products among the trendsetting elite and fashionable of São Paulo and Rio de Janeiro in order to stimulate an eventual nationwide trend.

In another sector, in October 2005 the Portuguese author José Saramago, a Nobel laureate, launched Brazil's first FSC-certified book, made entirely from certified paper and cardboard. Similar tactical efforts to increase demand from government purchasers have also been successful. Following a lengthy campaign by Greenpeace, the state of São Paulo has agreed to give preference to certified wood in its procurement decisions. Considering the size of the state's budget for construction, once implemented this should prove a major boost to domestic demand for certified wood.

Suppliers of commodity wood products, however, report little demand for certified wood. Retailers in Brazil are uninterested in offering certified wood to a public that is largely unaware of the FSC program or what certification means.[36] Brazil's high number of chain-of-custody certifications reflects export activity more than domestic demand for certified products. Still, several managers interviewed expressed hope that, as other, more trendy products spread the news about forest certification and wood labels, a domestic market for certified wood will develop.

Nonmarket Factors Affecting Demand for Certification in Brazil

Producers in Brazil report few direct nonmarket pressures in terms of NGO campaigns or threats of legal action that affect their decisions regarding FSC certification.[37] This seems surprising, considering that Brazil has a sizable and diverse activist community and that environmental issues have received so much attention in the country. Part of the explanation is that deforestation in Brazil is due mostly to illegal logging and the clearing of land for agriculture, not to formal forestry operations. The main thrust of local activists and regulatory campaigns to stop deforestation target illegal operations and/or corrupt regulators but have little effect on formal forestry operations.

Also, though Brazil's environmental laws are relatively strict, in practice the courts are overloaded, slow, and readily corruptible. Federal prosecutors and judges sometimes act aggressively against alleged environmental violations, but this is more common in urban than in rural areas, where wealthy landowners are often important political players on the local scene. In the Amazonian basin, local regulators and judges frequently have personal political or financial incentives to overlook activities that enrich local politicians, their friends, and families.

This is not to suggest, however, that Brazilian civil society has had no influence on FSC effectiveness. Widespread public concern and activism over deforestation and illegal activity in the nation's rainforests is a critical contextual factor that helps explain the industry's interest in forest certification. Forty years of highly publicized crisis in Amazonia has had a major impact. Widespread social awareness of the plight of the nation's forests has shaped the attitude and behavior of Brazilian forest industry officials, even in the absence of direct pressures from activists.

This accounts for the fact that, for the majority of FSC certification holders, certification followed a previous commitment to high standards of environmental management and practice. The poor reputation of logging and forest management industries in Brazil has resulted in a clear division between large-scale, legal operators who generally implement high standards of practice, and smaller-scale, illegal operations bent only on short-term profits. Few are the formal, legal producers who operate at levels of minimum legal compliance or skirt the law, as in other industries. As a result, for Brazilian forestry operators, obtaining certification tends to be either relatively easy or unthinkable.

The government in Brazil, at both the federal and local levels, has had little direct influence over demand for FSC certification or certified products. São Paulo state's promise on procurement policies is thus far the nation's only example of a state-based program promoting the purchase or use of certified wood. Forest managers report no regulatory benefits from or state recognition of the value of FSC certification.

However, while state agencies fail to *directly* influence the FSC in Brazil, the longstanding ineffectiveness of government regulation, and the resulting public skepticism, have *indirectly* promoted the program's growth. The fecklessness of state regulators, along with widespread public concern over deforestation, have generated significant public demand for independent, private verification of good practice. One of the first firms to certify with FSC did so because its managers sought an independent source of verification and recognition for what they believed were excellent environmental practices. Company officials wanted an entity that "was not from the government and was not from the industry. Someone we were not paying, that was checking our operations independently."[38]

In the coming years the government's role as a source of demand for forest certification may grow. The recently passed federal Public Forest Management Program aims to require independent sustainable management certification for all economic activities in public forests not set aside for conservation—approximately one-third of the Amazonian region. While this certification is not required to be through FSC, many believe that the law, as it is applied, will push the vast majority of forest managers in that direction. Beyond this, the only existing example of state action in support of demand for FSC comes from the state of São Paulo.

SUPPLY SIDE

Transnational Firms and NGOs in Argentina
The FSC chapter in Argentina has depended completely on patronage from transnational NGOs, particularly the World Wildlife Fund (WWF). FSC-Argentina is coordinated by and housed within the Fundación Vida Silvestre, the Argentine partner of the World Wildlife Fund. As of 2006, FSC-Argentina is coordinating six different standards-writing initiatives, five regarding regional native forests and the sixth for a standard on plantations, as well as pursuing public outreach. All of these initiatives are

administered by a single person who shares time between the Fundación and employment as a university professor.

Environmental and social rights NGOs in Argentina tend to belong to particular regions, municipalities, or communities and to focus in their operations on local issues. This complicates the efforts of FSC administrators. For the national standards initiative, these various regional organizations, some of which are quite small, must be convened regularly in order to participate. For regional initiatives, the FSC must piece together local-level consensus each time with new actors and organizations. The few nationally active environmental NGOs in Argentina are largely focused on urban pollution, water quality, and other issues that threaten the country's mostly urban population, and forest management is not considered a priority. Greenpeace, for example, supports the FSC but in Argentina focuses its energies on trying to protect native forests in the northwest of the country, a situation where there is no market demand and certification has little relevance.[39]

Transnational Firms and NGOs in Brazil

> Brazilians are like monkeys. We're excellent mimics. We will adopt anything, any foreign model, *immediately*, if we think it will work for us.[40]

Similar to the Argentine case, in its early years (1996–2002), the Forest Stewardship Council chapter in Brazil was supported almost entirely by money, staff, and office space provided by the World Wildlife Fund. That, however, is the extent to which foreign or global NGOs have led Brazil's FSC chapter. Although the WWF still supports FSC-Brazil, the national chapter is entirely independent. FSC-Brazil has its own headquarters in Brasilia and employs a full-time staff of six.

The key reason why the FSC has flourished in Brazil—unlike in Argentina—was that before it became available via FSC-International, there existed already in Brazil a diverse coalition of actors interested in forestry certification for their own reasons. Because the Brazilian forestry sector was already looking for the means of legitimizing itself, it was not only environmentalists and forest community representatives that sought to adopt the FSC. Firms and industry associations had a vested interest from the beginning.

Forest certification through local service providers and supported by the NGO Rainforest Alliance, among others, was available even before the FSC was established in 1994. Government-sponsored efforts to write an official national standard, which later evolved into the CERFLOR system, also predate the FSC. Brazilian certifiers and industry experts who had experience with these programs participated actively in the original meetings in Canada and the United States that led to the formation of the FSC.

Already in agreement with the objective and model of sustainable forestry certification, these individuals and the NGOs and firms they represented embraced the FSC project, participated actively in the Working Group formed in 1996, and promoted the FSC initiative among their colleagues. The core of this network consisted of an informal group of Brazilian experts with experience working in the forestry industry and with international certification agencies. Many of these individuals were working for major forestry firms or had worked in the industry previously. Their participation helped build interest in the project and provided legitimacy in the eyes of industry actors to what is commonly perceived around the world as an NGO-led regime.[41]

This large, well-connected, and active local group of Brazilian and foreign advocates and coordinators provided several benefits to the Brazilian FSC initiative. The fact that the Working Group was dominated by Brazilians, and included several Brazilian NGOs, muted criticism of the FSC as foreign interference.[42] Also, in a move different from the approach in Argentina, the Brazilian Working Group selected as its director the CEO of a major domestic forestry company. This signaled that although the program was founded primarily by NGOs, it was not incompatible with industry competitiveness or growth. This director was able to contact his colleagues directly and solicit their participation, and most of the country's major industry players contributed to the process of defining local FSC standards. This effectively isolated conservative opponents, and the Brazilian FSC chapter was not challenged by any lasting ideological split among major producers regarding the value of forest management certification or sustainable practice, as is the case in Argentina.[43] This result was not achieved through serendipity. The Working Group acted strategically to avoid collective industry opposition.

> The native forest sector is not organized, so we faced no strong opposition from them. The only possible source of opposition there was from the Association for the Export of Tropical Lumber . . . so we put them on the economic council of the FSC. We mapped who could be a potential adversary, and targeted them to bring them into the group. We did not give a spot to any company in the Amazon. Instead we gave one of six seats on the Economic Council to the Association. We did the same thing with the pulp and paper industry. It negated any resistance within the industry, it blocked any notion that the process favored one company over another, and it gave these industries control over their participation.[44]

In recent years, this network of FSC participants and supporters has been responsible for many of the activities that have made the program effective. Social and environmental NGOs and certification agencies play various critical roles in support of the FSC. NGOs call to public and government attention activities they view as injurious to local communities or the environment, presenting risk to firms and operators who ignore standards or laws. They also publicize certifications and commend operators who act in line with standards of sustainability. Along with certification providers, NGOs spread information about the certification of forests, mills, and other facilities, offer technical and financial assistance, participate in FSC standards negotiations, coordinate participation by their clients or local partners, and monitor the activities of certified producers to report any noncompliance.[45] In this way Brazil's nationwide network of local and international environmental and community rights NGOs play various roles that give strength and credibility to the FSC program.

This network consists of dozens of individual organizations with their own interests, orientations, priorities, and prerogatives. The agendas of public interest NGOs frequently contrast with those of industry representatives, but divisions also exist within these groups. The dominant organizations in the FSC (in Brazil as in most national chapters) are offices of global environmental NGOs including the World Wildlife Fund and Greenpeace. Other key players, however, are more radical and aggressive in their approach to protecting forests, or more committed to the concept of sustainable exploitation rather than conservation.

The dependence of the FSC as an organization on the wide-ranging support of these various entities is sometimes a cause of concern. When promoting the program to companies, communities, and government agencies, these organizations do not always use the language or arguments that FSC staff would prefer.[46] In another example, monitoring of compliance with FSC standards is transparent and open to anyone. Recently Greenpeace has been particularly active in questioning the behavior of a few certified operations. While it is free to do so, this action on the part of what is perceived as a particularly aggressive and publicity-seeking NGO makes some FSC-certified firms uncomfortable.[47] Also recently, politically organized landless or squatters' movements have also become increasingly aggressive in their actions and often target FSC-certified forests for their squatting because their managers are seen to be especially vulnerable to civil protest.

One of the program's current challenges is to establish more clearly its independence from the agendas of these various organizations, whose primary objectives and tactics may differ from the FSC's official position of neutrality.[48] Industry officials warn that groups from the "radical Left," particularly some social rights NGOs and the "Green Desert" antiplantations campaign, have gained too much influence within FSC.[49] These groups, particularly landless and squatters' movements, have been especially aggressive under the Lula presidency, because their poor members are key supporters of the ruling Workers' Party. Industrial forestry operators watch warily for indications of shift within the governing bodies of the program both nationally and internationally. Plantation managers especially warn that a perceived trend toward radicalism, coupled with controversies at FSC-International over the certification of tree plantations in general, may drive firms toward alternative certification regimes such as the government-sponsored CERFLOR.

The Role of State Actors in Argentina

The Argentine federal government is divided in its attitude toward the FSC. The Ministry of the Environment, responsible for environmental conservation, supports the FSC as a standard for managing native forests. This ministry, however, has little resources or political leverage. The more powerful Secretary of Agriculture, Ranching, Fishing, and Food (SAGPyA), which oversees policies regarding planted forests, announced in 2004 that it would begin composing a national standard for sustainable forest management to cover plantations. This initiative, cosponsored by the European Union, aims to provide market benefits for exporters similar to those of the FSC, but under requirements more sensitive to the concerns of the national industry.

Argentina's national institute for standardization (*Instituto Argentino de Normalización y Certificación* [IRAM]), commissioned by law to write national technical and operational standards, has also begun to write a national forestry standard. These two initiatives vary in two significant ways. The Secretary of Agriculture's program, modeled after Chile's CERTFOR system, intends to base its certification on flexible system standards akin to those of the EU's Programme for the Endorsement of Forest Certification (PEFC). IRAM, on the other hand, aims to write performance-based standards like those of the FSC, and would prefer mutual recognition between its national standard and those of the FSC.[50]

Industry officials give short shrift to these state-based standards programs. Several officials are dismayed at the intragovernmental competition to write a national standard. IRAM is widely considered the appropriate institution for the writing of the standard. The Secretary of Agriculture's program was formed around the funding the Secretariat was to receive from the Europeans rather than any logical basis for its involvement. Some observers suggested that the SAGPyA is proceeding independently with the support of the major TNCs, who view IRAM as too independent.

> I welcome [government agencies] to the certification model. If they could create a certification program that would improve forest management and get the backing of the industry groups, that would be wonderful. But today at least, and in this country, at least, if you think you can go to the market with a logo that is backed by the government of Argentina and by

the largest companies in Argentina, and sell it next to another logo that is backed by a coalition of independent international organizations, and the consumers will see these as equal, you are completely totally crazy.[51]

Many also question the capacity of the federal bureaucracy to complete the project, due to a legacy of poorly conceived and uncompleted policy initiatives toward the forestry sector.

When talking about government programs, especially for small companies, remember that everything now is viewed in context of Law 25,080. Lots of people, and not just the large companies, lost a lot of money [because the compensations promised in that Law were never paid]. Producers are very angry with the government, for failing to keep its promises. So these days no one will listen to the government talk about more financial support for this or that. These companies are focused on the money they're already owed. They're not going to involve themselves with anything else that the government supports.[52]

As one official phrased it: "In this country, the label 'governmental' condemns any initiative or plan to failure. It is a death sentence."[53]

The Role of State Actors in Brazil

The federal Environmental Ministry (*Ministerio do Meioambiente*, or MMA) officially supports voluntary forest certification. However, the MMA and especially its enforcement branch, which is run by the *Instituto Brasileiro do Meio Ambiente e dos Recursos Naturais Renováveis* (IBAMA), suffers from internal division and corruption.[54] Despite the ministry's official position, local IBAMA regulators often view FSC standards and audits as infringements on their authority. Some have recently withheld harvesting licenses to FSC-certified forest owners, allegedly because these firms do not pay the usual bribes because certification requires transparent accounting. Some of these certified firms have halted operations and fear that this may force them to sell out, most likely to a less scrupulous manager.[55] At least one firm recently went to court to pressure the agency to allow it to operate.[56]

As in Argentina, in the early 1990s the Brazilian forestry industry collaborated with the national standards agency (*Instituto Nacional de Metrologia, Normalização e Qualidade Industrial* [INMETRO]) to write an alternative national forest management standard for tree plantations. This *Certificação Floresta* (CERFLOR) standard is modeled on the U.S. industry's Sustainable Forestry Initiative (SFI). These systems are distinguished from the FSC in the flexibility of their standards and the weight of industry representation in their directorships. In the case of the CERFLOR, many industry experts view it as generally similar to the FSC, especially in the rigor of its environmental performance standards.[57] There are, however, two important differences. First, CERFLOR accepts transgenic trees while the FSC does not, which may in the future reduce the appeal of the FSC system.[58] Second, CERFLOR's standards are less exacting in terms of relations with local communities and open, public negotiations to resolve conflicts over land use or ownership.

FSC requirements regarding community relations pose a serious complication for a few major plantation operators that are engaged in prolonged disputes with indigenous groups over land rights and ownership. Such conflicts are spreading across Brazil. In the Amazonian region, squatters and land rights activists have begun to target certified forests for illegal invasion, since according to FSC standards the owners must engage in open negotiations and seek mutually beneficial solutions instead of offering the traditional response: expulsion at the end of a gun.[59]

Certification under CERFLOR standards has been an option since 2002. To date, however, only two forests have obtained CERFLOR certification. Despite its technical soundness and official sanction, the CERFLOR program suffers from a lack of market recognition and a general view that it is compromised by its close ties to major firms, a national industry association, and the government. While market demand for FSC-certified goods may be limited, industry officials and experts report that they have never heard of a client, domestic or foreign, demanding a logo under the PEFC system, with which CERFLOR is associated. As is the case in Argentina, state support for the program results in low credibility. In addition, CERFLOR faces an uphill struggle to gain participants because FSC has already achieved such a significant market position.

As is the case with demand-side action by the government, future influence may come more at the state than the federal level. As of 2006, three state governments in the Amazon region (Amapá, Amazonas, and Pará) sponsor incentives programs that support community-based forest certification through FSC by offering technical and financial assistance.

EXPLAINING THE GAP IN EFFECTIVENESS

Brazil's FSC chapter has outperformed its Argentine counterpart in both scope and strength. Both national chapters cover the same practice areas as defined by the global regime. However, the Brazilian chapter features five national standards that are more specific to local conditions, while Argentina's chapter has yet to complete its first, covering plantation operations only. In terms of strength, Argentina's FSC chapter has weakened in recent years, as the process for defining forest standards has floundered, the chapter has lost its official status with FSC-International. Monitoring and enforcement are provided from outside. In Brazil, the greater number of watchdog NGOs and a more active advocacy community provide greater oversight over environmental practice—and malpractice—in general, including that of FSC participant companies.

In terms of participation, Brazilian firms, forest owners, and forest managers do so in greater numbers and their certification totals have grown steadily over ten years. Participation in FSC-Brazil is concentrated, however, in the more developed southeast of the country and to a lesser extent in certain states in the north. The plantations industry and native forest managers, and their partners, are represented evenly. In contrast, Argentina's FSC forest certifications pertain mostly to smaller or medium-scale plantation operators, its chain-of-custody certifications are limited to the niche sector of higher-end wood products manufacturing firms, and there are no certifications for the cultivation of products from native forests.

The Brazilian national initiative has succeeded in building and maintaining an impressively robust, supportive coalition of environmental and community NGOs, industry actors, and, to a lesser extent, retailers of wood products. Brazil's FSC chapter had an advantage in terms of preexisting local interest in sustainable forestry practices, labeling, and certifications, the result of international scrutiny of deforestation in the country since the 1970s. In contrast, the various standards-writing processes in Argentina are managed by a single NGO that struggles, alone, to bridge various sectoral and regional divisions.

There is less difference between the impacts of these national chapters on member company practices. Industry officials and auditors in both countries attest to the stringency of FSC requirements, the seriousness of the auditing and verification processes, and the often high costs of certification. Brazil has locally established sets of forestry standards, while Argentina's participants must comply with universal, generic standards written by FSC-International. Nevertheless, the two are substantially the same because FSC-International demands adherence to its global principles.

As Chapter 2 explained, four factors in particular are believed to determine the effectiveness of the national chapter of a private environmental regime. These are market demands, transnational firms and NGOs, state policies, and the structure of the relevant industry. At this point, we can assess the relevance of these factors in causing the observed differences in the effectiveness of the FSC regime in Argentina and Brazil.

In these cases, there is little evidence that market demand has played a significant role. On the surface, market demand for FSC-labeled goods seems an important factor. Industry officials and experts in both countries state that firms will not participate over the long run unless there are clear benefits in terms of market share and/or profits. Also, managers in both countries report certifying under FSC standards because they hope to gain greater access to European and North American markets, or to obtain a higher price for their goods. Hardly any, however, report achieving any such benefits, and those that do operate mostly in small export sectors such as tropical hardwood exports, fiberboard, or niche goods. While some industry officials in both nations believe that certification is becoming increasingly important in achieving access to European wood and wood products markets, others are less sure. In Argentina in particular, managers tend to be taking a more wait-and-see approach, unconvinced thus far that FSC certification is worth the costs. FSC participants in both countries tend to be pessimistic about the prospects for significant price premiums.

The lack of market benefits for leading products such as commodity wood, pulp, plywood, paper, or cardboard products is commonly cited as a serious shortcoming of the FSC system. However, it is intriguing that no one in the industry or within the FSC in either country reports any firm or private owner quitting certification because of the lack of these realized benefits. If market-based returns are indeed a critical source of influence over managers' decisions to participate, and therefore over regime effectiveness, then clearly managers are willing to wait for years without any such benefits before they quit the regime.

These two cases indicate, instead, that companies tend to certify with hopes of obtaining market benefits, but that once they have achieved compliance they recognize other, less anticipated benefits. These additional benefits are sufficient for them to continue their participation, at significant cost, year after year, whether or not market-based benefits are ever realized. In addition, the logic of sunk costs may affect their decisions, particularly for smaller operators.

Another indication that market benefits are not driving participation is that these regime chapters' effectiveness does not correlate, at the sector level, with realized market benefits. Operators in sectors where participation rates are the highest claim no more market benefits than those in other sectors. Brazilian forest plantation managers, for example, report no market benefits whatsoever, and Argentine exporters of refined wood products have mixed experiences. Despite relatively significant international market benefits for certified exporters of tropical wood, growth in that sector has been slower than among plantations.

In both countries, firms tend to participate in the FSC because of its non-market benefits. The process of achieving FSC certification demands major improvements in management and forestry techniques, including in most cases the introduction of new management and accounting systems and technologies. Certification signals to peers and stakeholders that the firm is a serious and reliable partner, committed to excellence in its environmental and management practices. Furthermore, in Brazil certification helps to distinguish forestry companies that operate legally and responsibly from the thousands of others whose activities contribute to the national disgrace of uncontrolled deforestation. This suggests that the pressure for *social license* is extremely powerful in Brazil, where the national public seeks an answer to the problem of rampant deforestation.[60] In Argentina, pressures for social license to operate in an environmentally responsible manner are relatively low and less relevant to most firms than are the pressures for economic or regulatory license.

These cases demonstrate that although transnational actors played critical roles by introducing these regimes in Argentina and Brazil, the effectiveness of these national chapters has depended on the capacities, interests, and coordination of local actors. The success of Brazil's FSC chapter is due largely to the network of local individuals, firms, NGOs, and certification agencies that predated the FSC but found in the regime a useful tool for the pursuit of their common interests. Brazil's chapter is made up largely of Brazilian organizations, and virtually all of its key players are Brazilians.

In Argentina, transnational firms and NGOs have played an ambivalent role. On one hand, the country's FSC chapter is sustained by the support it receives from the massive transnational NGO the World Wildlife Fund. However, the chapter's strongest opposition comes from transnational firms, a group led by Chilean companies but including U.S. and European firms that supports efforts by the Ministry of Agriculture to create a more industry-friendly alternative.

In regard to the influence of state actors, findings from these cases are mixed. In terms of state actors as agents, neither case supports the view that effective local governance is necessary to bestow public legitimacy on private standards. Government agencies in Argentina and Brazil have thus far had little or no direct influence over firms' decisions whether to participate. These governments are not significant buyers of certified wood (although São Paulo state is considering doing so). They have not sought FSC certification for public forests, nor have they changed their regulatory approach to recognize FSC-certified operators. The fact that the state has little involvement in these forestry industries, neither as a regulator nor as a provider of services, carries over to certification. Even in cases where state actors seek actively to influence FSC participation by offering alternative state-supported programs, thus far they have been ineffectual. Particularly in Argentina, the state lacks credibility as a reliable partner of business. Indeed, one key source of the FSC's credibility in both nations is its unquestioned independence from the influence of local governments.

Nevertheless, state actors are important elements of the market and nonmarket structures that forest producers and consumers face in these countries. The success of Brazil's FSC chapter is in large part due to demand on the part of both firms and environmental activists for some credible mechanism to distinguish responsible forestry practices. This demand is the result of demonstrated inability of the Brazilian government, across decades, to control events in its Amazon territory. Brazil's forestry and wood products industries are tainted by this national record of ineptitude and corruption. As a result, certification by an international, independent organization like the FSC is required in order to do business internationally. Foreign buyers wary of involving themselves with illegally or unethically sourced wood cannot tell whether some stamp of government approval was earned or bought, and so these firms must look to independent sources to verify their good practices.

In this way, the legacy of decades of state ineffectiveness or indifference has had a strong positive influence on FSC effectiveness, though not in any way the government would have wished. Government actions and policies are indeed important, but only indirectly, as part of the incentives structure in which firms, NGOs, and other private actors operate. The importance of state actors derives from their ineffectiveness, which increases demand on the part of firms and other stakeholders for private forms of environmental regulation.

In contrast with the common wisdom that demand-side factors drive regime effectiveness, the cases demonstrate that in fact supply-side factors are much more determinative of regime success. In Brazil and Argentina it is capacity and organization on the supply side that mostly have determined the effectiveness of national FSC initiatives. Chief among these factors are the availability of partnerships among capable local partners and the resources and strategies of chapter administrators. Brazil's network of nationally active forestry-oriented organizations differs dramatically from the case in Argentina, where environmental NGOs and industry groups are more sharply divided across regions. The strategy of Brazil's FSC administrators to target, early on, the inclusion of key firms and industry actors successfully undercut ideological opposition from industry actors, a problem that remains critical in Argentina.

There is little evidence that degrees of industry concentration have had any systematic effect on these chapters' effectiveness. The Brazilian forestry sector is heterogeneous and diffuse, yet the FSC has been more effective there than in Argentina, where high industry concentration leaves its chapter vulnerable to the opposition of two major firms.

To conclude, these FSC case studies reveal the importance of local conditions and factors, in particular the configuration of interests among industry and stakeholder actors, and the strategies and capacities of local administrators. Other factors purported to be key determinants of regime effectiveness—advocacy by transnational firms or NGOs, and state actors—played relatively insignificant roles. Market demand remains important as an ultimate objective, but failure to realize significant market benefits thus far has had little effect. A more important source of demand, evident in Brazil more than in Argentina, is the legacy effect of decades of state indifference or failure at managing the nation's forests. As a result, even forestry or wood firms that operate far from any area of current deforestation are interested in the FSC as a tool to differentiate themselves from their many illegal and irresponsible competitors.

Charts 4.2 and 4.3 summarize these case studies. Two points are particularly significant because they fail to support claims common in the literature on global private regulation. First, transnational diffusion via trade or supply chain pressures is surprisingly weak in these cases. Second, state actors in these countries have played largely insignificant roles as agents pursuing their preferences regarding the expansion or restriction of the FSC. However, the legacy of longstanding state policies or failed policies leaving forests largely unmanaged is a structural condition with important effects for regime development.

An analysis of the importance of these findings within the larger picture of global private regulation is left to Chapter 7. Before that, it is useful to compare these accounts of the effectiveness of the Forest Stewardship Council in Argentina and Brazil with those of a contrasting, industry-administered private regime within a sector wholly different from forestry and wood products: the chemical industry's Responsible Care.

Chart 4.2   Summary of the FSC-Argentina case

Demand-side factors affecting the FSC chapter	Demand in European and North American markets drives interest in FSC certification.
	No domestic market or demand for certified products.
	No pressure across production chains. Large companies are not certified and certified chains of custody (from forest to mill to manufacturer to retailer) are not yet established nationally.
	Nonmarket pressures from regulators, advocacy groups, or public are negligible.
Supply-side factors affecting the FSC chapter	Administrative agency is capable, although the management of multiple initiatives across a large country stretches its few resources and small staff.
	No other NGOs or independent groups play significant supporting role.
	State policies toward FSC are divided; however, serious efforts are under way to create a state-sanctioned alternative to FSC's standards for plantations. The government's overall lack of credibility, however, hampers these efforts.

Chart 4.3   Summary of the FSC-Brazil case:

Demand-side factors affecting FSC in Brazil	Market demand in Europe drives interest in FSC certification, especially for tropical wood exports. In other sectors, and in the domestic market, there is little demand and no price premium for certified products.
	Certified chains of custody aimed chiefly at export markets are well developed, but thus far there is little market return, domestic or foreign.
	Direct nonmarket pressures from regulators, advocacy groups, or the public are negligible. However, legacy of failed forest management in Brazil has created a climate in which legal, formal forestry operations face high public and commercial scrutiny. FSC certification is viewed as a useful tool to demonstrate corporate responsibility and good management.
Supply-side factors affecting FSC in Brazil	Administrative agency is capable, with a full-time staff of six. However, organizational capacity is strengthened most by a network of active NGOs and certification agencies that share the goals of the FSC.
	Brazilian chapter successfully engaged industry leaders from the start, isolating ideological opponents and linking forest certification with successful, well-managed businesses.
	Federal and state governments officially encourage forest certification, but this is not yet true in practice. In the Amazonian region corruption is rampant among regulators, who sometimes penalize certified operations.
	The national standards agency offers a state-sanctioned domestic alternative to the FSC. However, interest in this program has been limited by its poor credibility and lack of recognition from the market.

CHAPTER 5

THE INTERNATIONAL CHEMICALS MANUFACTURING INDUSTRY AND RESPONSIBLE CARE

Chemicals manufacturing is the bedrock industry of a modern industrial economy. The chemicals industry changes raw materials from nature such as oil, grains, natural gas, and wood into the products that define contemporary life, from soap to clothing, food to cars. Every other major industry depends upon chemicals as inputs, both in the products they make and the processes for making them. In 2006, global chemicals sales totaled over US$2.2 trillion. Only the food industry is larger.

The chemicals industry is vast, diverse, and complex in its organization. Chemicals are used in the design, production, packaging, and preservation of virtually all consumer goods. This diversity complicates attempts to delimit the industry as a whole. Petrochemicals, plastics, agrochemicals, and pharmaceuticals, for example, are major industries of their own. The chemicals industry encompasses all of these and several others, as well as the shared input lines (fuels, basic chemicals, etc.), processes, and technologies that these various sectors require. As we shall see, this diversity and complexity complicates the establishment and administration of industry-wide regulatory regimes.

Despite this diversity, the global industry is extremely concentrated. Tens of thousands of producers operate worldwide, many in developing countries out of garages and kitchens. Nevertheless, the world's two dozen leading chemicals manufacturing firms dominate global production, sales, research, and product development. Firms such as BASF, Dow Chemical, Bayer, and

DuPont are among the world's largest, most diverse, integrated, and consistently profitable companies. They are also extremely active in national and international policymaking in areas that affect their operations, including the environment.

The production of chemicals has, for decades, been concentrated in the industrialized countries of Western Europe and the United States. However, over recent years the industry has boomed in Asia and other industrializing regions, reflecting the overall trend in global trade and investment. Because chemicals are necessary for all other major manufacturing industries (including defense), governments keen to industrialize and improve their self-reliance have sought to create and grow their local chemicals industries via subsidies, market protections, tax breaks, and strategic state-led investment. Tactics that worked, for example, for Germany and the United States in the late 1800s have been followed since the 1950s in India, Brazil, Argentina, and China.

Competition in the chemicals industry is driven by access to inputs (particularly petroleum and natural gas), proximity to major markets, and opportunities for integrated production.[1] By facilitating and reducing the costs of the transfer of products, money, and technology across borders and oceans, the liberalization of international trade has transformed the industry. Over the last 20 years, companies have established major new plants throughout the developing world, both to service these growing markets and to take advantage of their human and material resources. Also, heightened competition and deregulation in the 1990s sparked a rash of mergers at the global and national levels, including in developing states where formerly state-owned companies were bought up by industry leaders, including in many cases transnational firms.

As with forests and forestry, issues regarding the manufacturing, use, and disposal of chemicals have long been at the center of environmental politics. For decades, environmentalists and public health advocates have waged an aggressive campaign for greater state controls over the production, use, and disposal of thousands of toxic and hazardous chemical products. Rachel Carson's book *Silent Spring*, which in 1962 galvanized the modern environmental movement in the United States, documented the pollution and sickness caused by the extensive use of dichloro-diphenyl-trichloroethane, or DDT, and other agro-chemicals. This issue, combined with rising alarm over deadly outbreaks of mercury poisoning in Japan, as well as mercury-caused illnesses in North America and Europe, led governments to convene the 1972 UN Stockholm Conference on the Human Environment. The Stockholm Conference produced the UN Environment Programme and a UN International Register on Potentially Toxic Chemicals, and marked the beginning of an ongoing process of stiffening rules and standards regarding industrial pollution.

The dominant position of North American and European firms in the global industry has shaped the making of rules in this issue area. Early on, the Organisation for Economic Co-operation and Development (OECD) took the initiative in writing international standards for the production and trade of toxic chemicals. Major chemical firms pushed their governments to take the reins in these issues instead of leaving them to the more open, unpredictable forum of the UN. The OECD first met on toxic chemicals in 1966 under the title Expert Meeting on Research on the *Unintended Occurrence* of Pesticides in the Environment (author's italics). Throughout the 1970s the OECD worked to establish a chemicals testing program, which in the early 1980s became de facto rules for chemicals testing, and it later established similar global standards for good laboratory practice. These important steps toward the global monitoring and control of toxic chemicals were achieved through close collaboration between the world's largest chemical manufacturers and their home state governments.

Chemical corporations have had more success at influencing regulation at the international level than domestically. In the United States the book *Silent Spring* shocked the nation, and the chemical industry's concerted and prolonged negative campaign against its author, publisher, and message did not reassure the public. Instead, the chemicals industry became synonymous with toxic pollution, misinformation, and the abuse of public trust. The tightening of controls over chemicals, their production, labeling, use, and disposal, have been major objectives of many American environmental groups, which they have pursued successfully in various ways, including via public campaigns, lobbying, and aggressive litigation.

Throughout the 1970s the chemicals industry fought back against demands for tighter regulation and sought to allay public concerns through public relations and informational campaigns, but by the early 1980s they were losing ground. Environmental movements in the United States and Europe were increasing in their numbers of members, their range of activities, and the aggressiveness of their tactics. The public conflated toxic pollution and emissions with new fears over global warming and the depletion of the atmosphere's ozone layer, and denials on the part of the chemicals industry and others lost credibility. Across the world, governments responded with new laws, stricter standards, and regulatory agencies for the control of toxic and hazardous chemicals.

In 1984 an emissions leak at a Union Carbide plant in Bhopal, India, killed over 3000 workers and local residents. This catastrophe sparked international fury against the chemicals industry and alarm over the exploitation of communities, workers, and the environment in poor countries where laws and state agencies were weak and easily corrupted. Governments around the world passed new safety and emergency preparedness laws and regulations, many of which targeted the chemicals industry in particular.

One prominent example was the creation in 1987 of the U.S. Toxics Release Inventory (TRI), a national database of toxic chemical waste and emissions based on data required of all manufacturers. Compared to tightening state controls in Europe, the U.S. government sought a more business-friendly, open-information approach to regulating toxic chemicals. The TRI enhanced the public's access to information on the use, storage, and release of toxic chemicals in their own neighborhoods, and by doing so moved the country down the road toward more information-based and market-centered regulation, instead of command-and-control systems based on government monitoring and action. By exposing the performances of individual companies, the TRI also widened the political divide in the industry between firms that sought to demonstrate a commitment to emissions reduction and responsible waste management and those that resisted change.

The global fallout from the Bhopal tragedy demonstrated to the industry that disasters and irresponsible behavior associated with any chemical firm anywhere contributes, in the public eye, to antagonism and mistrust toward the industry as a whole. By the mid-1980s it was also clear that the industry's aggressive public relations campaigns were not working, and in fact were counterproductive. Surveys conducted by the industry association and individual firms revealed that the public image of the chemicals industry, after a decade of informational campaigns and public relations, was extremely poor.[2] Worried about the long-term effects of public distrust, as well as the persistent tightening of regulations, leading chemicals firms began to push the industry to take a more proactive, responsive approach to the management of pollution and toxic substances.

THE RESPONSIBLE CARE INITIATIVE

Responsible Care was established in Canada in 1985, the result of discussions between the U.S. and Canadian chemical manufacturers' associations about the institutionalization of pollution prevention models across the industry. U.S. firms including Dow Chemical were its leading advocates, largely because the program encouraged the rest of the industry to implement environmental management programs similar to their own.[3] From the beginning, industry officials advocating Responsible Care described it not as a regulatory regime or initiative, but as an effort to encourage a different business culture in the chemicals industry.

Once established in Canada and the United States in the late 1980s, the International Council of Chemical Associations (ICCA), along with the major U.S. TNCs, promoted its establishment in other countries where they operated. By the early 1990s, Responsible Care was in place in countries around the world. The Argentine and Brazilian industry councils, for example, established chapters in 1992 and 1993 after advocacy on the part of DuPont and Dow Chemical. Currently, Responsible Care operates in 53 countries and participating firms account for nearly 90 percent of total global chemical production.[4]

Responsible Care (RC) is the world's leading model of an industry-led private environmental regime. Its goal is to promote continuous improvement in the health, safety, and environmental management practices and performances of chemical producers and their partners. Sponsored and led globally by the ICCA, at the national level the regime is administered via accredited national industry associations. As with the decentralized management structure of the Forest Stewardship Council (FSC), this design is intended to provide the RC regime with flexibility before different local and national conditions and needs. Still, the global regime mandates that all national chapters feature several common components. These include sets of codes and indicators against which member firms' management or performance are evaluated, a global title and logo, procedures for communication with key stakeholders, and methods of verification of compliance with the terms of members' commitments. As with the FSC, the ICCA as global administrator defines the regime's global principles, guidelines, and general structure, but national chapters must themselves make local rules, write codes of practice, and implement verification and enforcement processes. All these functions, and any additional benefits, must be overseen, staffed, and funded—for the most part—by national industry associations.

The codes and indicators used by Responsible Care typically focus on management standards instead of specific performance targets, though the performances of member firms in basic areas like emissions and waste reduction are regularly measured and tabulated to reflect—administrators hope— continued improvement across the industry. The aim is to demand and verify continued improvement in practices, rather than to set specific goals or measure outcomes. As a result, Responsible Care does not ensure that every member firm or its products comply with any specific standards for environmental performance. For this reason, unlike the Forest Stewardship Council, RC provides no basis for product labeling. The use of the RC logo on products is prohibited by the ICCA.

Thus, Responsible Care does not offer consumer-oriented differentiation or price premiums, as the FSC aims to do. The regime does not offer any market-based advantages from participating. Instead, participating firms can claim membership in a club of companies committed, at least on paper, to extraordinary standards of environmental and safety management. Ideally, they also participate in regular meetings where information is shared on best practices and new technologies, and receive annual audits and advice on means of improvement. RC membership should also bestow some degree of reputational benefit, because participation should indicate to potential clients, regulators, and the public that the firm is committed to high quality, responsive management, and safe, clean operations. Indeed, one of the key advantages of participating in Responsible Care should be its use as a shield against negative scrutiny from outsiders.

The extent to which members obtain these benefits, however, is uncertain and varies from country to country. In most countries, clients and consumers are not familiar with RC, and its credibility is compromised by the fact that it is run by the industry itself. In most country cases, verifications of compliance with RC codes, as well as all performance evaluations, come from companies' self-reporting, and the auditing process is for the most part closed to outside participation or scrutiny.

Whereas the Forest Stewardship Council exemplifies third-party regulation via external, independent actors, Responsible Care is a case of second-party self-regulation, or regulation through a collective of a company's peers within the industry.[5] Facing the trade-off between control and credibility, Responsible Care administrators and member firms opt for control. The regime responds first to the interests of industry managers who wish to engage in environmental management but without entailing undue costs, and secondarily to the need to placate outside critics.

Over time, the ICCA and many national chemicals industry councils have recognized the need to improve the credibility of RC by sanctioning noncompliant firms, and they have accepted some degree of independent oversight. In 1996, global administrators introduced verification processes as a basic requirement for all chapters in order to push beyond self-reporting. Beginning in the late 1990s, chapters in Europe and the United States began to encourage external audits of compliance with RC codes, under the model of ISO 14001 certification. Also, national chapters around the world began to make RC participation, and annual audits, a requirement of all chemical council members. In 2002, the American Chemicals Council made independent audits and certification of compliance with Responsible Care guidelines mandatory for all industry council members by 2005. Again, other nations including Brazil have followed suit. Some industry councils, including those in the United States and Brazil, have expelled members who did not comply with Responsible Care reporting requirements.

Responsible Care demands of its members continual improvement in their safety, health, and environmental management practices and encourages firms to have regular communication with their communities and other stakeholders regarding their policies. Yet member firms can set the terms of their own compliance by defining "continual improvement" favorably to their needs, and self-reporting allows members ample room for maneuvering around sensitive issues. So how effective is Responsible Care? Studies of RC in the United States indicate that the impact of participation varies dramatically across firms and across different elements of the regime. Firms that showed a previous commitment to environmental management before joining RC tended, under the RC system, to improve their practices. However, in the cases of firms with relatively poor environmental performance there was little correlation between membership and changes in practice. Furthermore, nonmember firms tend to improve their environmental performances at a faster rate than members, suggesting that RC, at least in its early years, served as a shield behind which many poorly performing firms could hide.[6]

RESPONSIBLE CARE IN LATIN AMERICA

Responsible Care chapters have been established in nine countries in Latin America. Mexico adopted the program in 1991, Argentina and Brazil in 1992, Chile and Colombia in 1994, Peru in 1996, Uruguay in 1998, Ecuador in 1999, and Venezuela in 2002. A summary glance at some key characteristics of these national RC chapters demonstrates important contrasts in their development and strength, differences that result from idiosyncrasies in the local industry.

Mexico's chapter is relatively advanced due to the close relationship between that country's industry and the United States and the leverage major U.S. firms have within the Mexican industry. Since 2002, the Mexican council has required third-party audits of compliance with RC guidelines. In contrast, Uruguay's RC chapter is administered by Argentina's industry council—a reflection of the small size of Uruguay's industry—and Venezuela's chapter requires no more than informal, annual self-assessments. Ecuador's national chapter is unusual in that its operations have been funded and partly managed by a major national environmental NGO.[7] Chapter 7 explores the cause of such differentiation, within the context of lessons learned from Argentina and Brazil.

Chart 5.1 compares Responsible Care chapters in four South American cases. Chile and Venezuela provide points of reference for the Argentine and Brazilian chapters, which are the focus of our study and the next chapter. Along with Mexico's chapter, these were established very early on, but their rates and degrees of development have been distinct. In terms of regime size, or participation rates, Venezuela's chapter clearly lags behind the others both in total number and in its share of the total industry. In the region, only Brazil's chapter mandates the participation of all chemical council members. Without this requirement, Brazil's participation numbers would likely be more in line with those of Argentina and Chile. In terms of strength, Brazil's chapter again exceeds those of its regional partners. Only in Brazil, Chile, and Ecuador do RC chapters allow any form of outside participation.[8]

Chart 5.1   Comparison of Responsible Care chapters in four South American countries

	Argentina	Brazil	Chile	Venezuela
Year established	1992	1992	1994	2002
Number of participating companies*	64 (70% of industry association members)	108 (100% of industry association members)	80 (80% of industry association members)	24 (20% of industry association members)
Verification system	Program has its own proprietary audit system, conducted by a government-linked certification agency. Results not publicly available.	Audits by an independent agency and with community participation are mandatory since 2004. System is complementary to ISO 14001 certification. Results not publicly available unless firm chooses.	Verification and audit system includes community representation, but is controlled by industry council. Results not publicly available.	No third-party verification. Indicators only defined in 2004.

* as of October 2005.

Sources: Responsible Care *Global Status Report 2005* (*www.icca-chem.org/pdf/icca005.pdf*).

As the next chapter will explain, Brazil's and Argentina's chemicals industries dwarf those of any other Latin American country except Mexico. Their similarity in size and importance to the national economy, coupled with the sharp dissimilarities in the size and strength of their national RC chapters, makes theirs an ideal pairing for cross-national comparison.

CHAPTER 6

RESPONSIBLE CARE
IN ARGENTINA
AND BRAZIL

> I tell you one thing. If there is a big industrial environmental disaster, a big accident or scandal, not only in Argentina but anywhere, they will look closely at the preventative system in place, all these audits and certifications, and these programs like Responsible Care, and it will be very bad for all of them.
>
> *Manager at a local subsidiary of a U.S. chemicals company,*
> *October 13, 2004*

The chemicals industry is founded on high technology and precision management, is globally integrated, and is driven in large part by a couple dozen transnational goliath such as Dow Chemical, BASF, and DuPont. The industry has been under intense pressure for decades to reduce toxic emissions, waste, and other negative by-products of chemicals use. With these characteristics, and with strong motivation, the industry—which also features highly competent collective associations at the global and national levels—should be capable of implementing a very successful global environmental regime. Indeed, its own publications and reports declare that it has. However, as with the Forest Stewardship Council (FSC), a close look at the operations, size, and strength of national regime chapters in developing countries paints a different picture.

This chapter applies the same analytical framework used in Chapter 4 to compare the effectiveness of the Responsible Care (RC) chapters within Argentina and Brazil. As Chapter 4 described, the FSC is run primarily by environmental and community and labor rights NGOs, and it defines its forestry standards locally through an open process of tripartite (industry, environmental, and community representation) negotiations. In contrast, Responsible Care exemplifies industry-led environmental regimes where outside stakeholders are either excluded entirely or granted limited access to the administration, procedures, and records of the regime.

Because of this key difference, these regimes seek legitimacy within local markets and communities via different strategies and coalitions. The FSC pursues legitimacy through convincing firms and private forest owners to participate in a regime championed by environmental progressives. RC tries to leverage technical and managerial expertise, and exercises frequent self-reporting, in order to gain the respect of outside observers including regulators and environmental groups (while shunning their direct involvement).

This chapter examines national RC chapters in Argentina and Brazil to reveal the degree to which factors at the national industry level determine the overall effectiveness of the RC regime. Similar to the cases of the FSC, these studies show the persistence of effects from past industrial policies, as well as the importance of attitudes among the public and managers within the industry about the environmental effects of chemicals manufacturing. Of special importance is the pattern by which particular national conditions lead, within the industry, to the formation of administrative agencies of drastically different capacities and effectiveness.

Before examining the industries themselves, however, and their experimentation with Responsible Care, it is essential to clarify the political-economic context within which they operate.

FROM STATE-LED GROWTH TO LIBERALIZATION: CHEMICALS PRODUCTION IN ARGENTINA AND BRAZIL

Under the military governments of the 1960s and 1970s, which supported and encouraged large-scale industrialization, the Brazilian and Argentine governments looked with favor on their chemicals manufacturing sectors. Like many governments, they viewed the domestic production of industrial chemicals as part of the base upon which would be built other strategic industries such as steel, construction, autos, and defense equipment. To boost the local chemicals industry, tariffs were raised on chemicals imports, and state-owned companies and financing agencies partnered with foreign firms to build local facilities to supply national markets.

Brazil did so most aggressively, as Peter Evans described in his 1979 study of the country's state-led industrialization model. Despite the country's oil and gas reserves, its state-owned firms lacked the technology and capital to refine its resources into second-generation products at a scale sufficient to support national growth. In 1972, a local industry group established the country's first major petrochemical production center in the state of São Paulo. Shortly after, the state-owned company *Petroquisa* headed a partnership among itself, local capital, and foreign companies to establish two more large-scale chemical "poles" in the northeastern state of Bahía (in 1978) and the southern state of Rio Grande do Sul (in 1982).[1] The Argentine government pursued a similar model of combining the efforts of the state, domestic industry leaders, and foreign firms (i.e., what Peter Evans termed the "triple alliance") in building a chemical industrial center outside of the port city of Bahía Blanca, in 1978.

These triple-alliance projects, and the macroeconomic policies of which they were a part, turned out very differently in Argentina and Brazil, and those differences have colored their national industries ever since. Both states used various means to protect local capital in the 1960s and 1970s, and kept local prices higher than global prices. However, Brazil's government was more successful at allowing domestic prices to decline over time, which encouraged local companies to merge and diversify into different sectors. Taking advantage of the attractiveness of its huge internal market, Brazil's incentives to foreign investors evolved over time to include national origin requirements and demands for the sharing of technology, which boosted the competitiveness of local producers. In contrast, the Argentine government was unable to build important secondary industries, partly because of the smaller size of its consumer and industrial markets. What began as a technocratic program in Buenos Aires evolved into a set of politically expedient subsidies and protections for local industry, shielding them from outside competition and stunting productivity.[2]

In both nations, these massive state protection and subsidization programs generated enormous debt. With the rise in global interest rates in the late 1970s and early 1980s, these debt loads proved unsustainable. During the "lost decade" of the 1980s these governments were forced to declare bankruptcy, renegotiate their debt payments to international lenders and commercial banks, and abandon the state-led development models that defined industrial policy in the 1960s and 1970s. Instead, they turned to neoliberal economic policies, particularly fiscal discipline and the opening of their economies to global markets. Across virtually every industry, governments lowered tariffs, reduced restrictions on foreign investment, privatized state-owned enterprises, and slashed regulations aimed at protecting local firms. As with most industries, this rapid deregulation amid economic recession jolted the chemicals manufacturing sector. Profits collapsed due to a sharp decline in global prices, combined with the lack of government protection to maintain higher prices at home.

Throughout the 1990s intensified competition, along with the lowered value of local assets, set off a wave of mergers and acquisitions across the region. Foreign firms in particular saw the opportunity to expand local production and increase efficiency in a sector that had long been protected by state policies. In Brazil, almost 60 percent of the industry workforce was cut between 1990 and 2001, while production increased. Brazil's chemical exports almost doubled in that period, reflecting the deeper insertion of the national industry into regional and global production chains.[3]

In both Argentina and Brazil, large chemical firms, including many transnationals, expanded into new product lines and forced many local firms into partnership or out of business. In response to the regional market integration established under the Mercosur agreement of 1991, many firms moved or concentrated their operations from a national to regional focus. Many focused their administrative and higher value-added operations in Brazil, where a cheaper *real* (compared to the Argentine *peso*, which was pegged to the dollar) made exports more competitive. The logic of regional integration oriented many Argentine producers toward the export of commodity products, largely to Brazil, instead of higher-end manufacturing.[4] The fact that Brazilian producers had, during the 1980s, developed large-scale capacity for second- and third-generation petrochemical products proved decisive in reshaping the regional industry into one characterized by core-periphery dynamics, with southeast Brazil—and particularly São Paulo state—as the new regional industrial core.

Argentina's economic and political collapse in 2001–2002 set that country's industry back further, though the years since have seen gains in profitability and investment in the sector in both countries. Once the Argentine economy stabilized in 2003, with the *peso* much weakened, conditions were favorable for growth driven by newly competitive exports, particularly in agriculture and food.

These national differences in the orientation and administration of state-led industrial policies in the 1970s and 1980s significantly influenced the capacities of these national industries to respond, in the 1990s, to liberalization and intensified foreign investment. In Brazil, the growth of major domestic chemicals producers during the 1970s and 1980s diluted the effect of industry "transnationalization" during the subsequent liberalization.[5] By the early 1990s, Brazil's largest chemical producers and their local supply networks were sophisticated enough to compete with, and in many cases form strategic partnerships with, foreign investors. By contrast, in Argentina, most local firms long protected from competition were unable to weather that country's rapid liberalization. Those that did so suffered tremendously under the crunch of economic turbulence and crisis during 1998–2002. As a result, by 2004 transnational firms and their subsidiaries dominated the local chemicals industry, especially in higher-end product lines.

Another residual effect of past industrial policies is that Brazil's chemicals industry (particularly its high value-added sectors) is extraordinarily adept at adopting practices and technologies from outside the region for their local operations. Whether due to the security of having a major consumer market to themselves, or to the success of previous state-backed projects in which local firms adopted technologies brought from outside, Brazil's chemicals industry seems to demonstrate an open, dynamic, and innovation-hungry culture. In contrast, domestic chemicals firms in Chile and Argentina tend to be more conservative in both their investments and their management style. Though only a generalization, this attitudinal difference is manifest in many aspects of the Brazilian industry's experience with Responsible Care.

ENVIRONMENTAL REGULATION AND THE CHEMICALS INDUSTRY

Argentina is not a compliance-oriented country.
Chief administrator of the program Cuidado Responsable del Medioambiente

Argentina and Brazil exemplify many developing countries in that serious environmental protections are written in the law and government policy, but they are not taken seriously. Legal implementation and enforcement suffer from bureaucratic confusion at the federal level, lack of technical application and enforcement at the state level, and potential corruption throughout. While Brazil's environmental regulatory climate is more institutionalized than Argentina's, particularly in the industrialized regions in the southeast, neither country enforces its environmental laws effectively nationwide.

Brazil's national environmental regulatory agency, the *Instituto Brasileiro do Meio Ambiente e dos Recursos Naturais Renováveis* (IBAMA), and several agencies at the state level are considered among the most capable in the developing world.[6] The nation's environmental legal framework is also robust and progressive. The national constitution of 1989 gives all citizens the right to a clean, healthy environment, and in 1998 the Congress passed a law raising penalties for environmental violations and allowing individual criminal liability. More recently, autonomous, federal public prosecutors given broad powers to defend the public interest have become key drivers of environmental enforcement. This impressive, multilayered regulatory system emerged in response to the international and domestic public concern over the health crisis at Cubatão in São Paulo state and the dwindling Amazon and Mata Atlántica rainforests.

On the other hand, the results of all the bureaucratic and legal attention that Brazilians have given to environmental issues have not been impressive. This is due partly to widespread bureaucratic inefficiency. Brazil's environmental regulatory regime is less hierarchic than that of the United States. Its effectiveness depends upon the coordination of efforts among institutions at different levels that are unaccountable to one another, which seldom occurs.

Also, even at the federal level where they are most institutionalized, environmental agencies lost political traction during the 1990s and, as a result, their funding decreased. Last, the general division of labor is that federal agencies and laws establish national priorities and policies, but state agencies are responsible for their implementation and enforcement. As is generally the case in Brazil, state governments vary dramatically in the effort and resources they dedicate to environmental regulation.[7]

Argentina's national environmental regulatory regime is even more complicated. Overlapping jurisdictional responsibilities, both vertically between federal and provincial agencies and horizontally among different federal or provincial units, generate confusion and inaction. Argentina created its first Environmental Secretariat in 1973, only to dissolve the agency in 1975. In 1991 President Menem revived the Secretariat and gave it political clout. Throughout the 1990s, constitutional reforms and various laws encompassed more stringent environmental planning and restrictions, but with little noticeable effect in practice.

As in Brazil, many of Argentina's environmental laws and requirements were imported from those of the U.S. Environmental Protection Agency or European laws, but have not been tailored in their application to suit local conditions. In 1999, President de la Rua weakened the Environmental Secretariat by placing it within the enormous Ministry of Social Development, and today the elements of federal environmental planning and regulation are spread across dozens of narrow federal agencies, including industrial development, agricultural research, forest management, and so on.

Also, the Argentine public is less organized and vocal regarding environmental issues than their neighbors to the north. In recent decades political, economic, and military crises have pummeled the country's civic and political culture. In the late 1990s and especially following the crisis of 2001–2002, problems of poverty, joblessness, and political uncertainty have overshadowed the nation's environmental woes. Even with increased public awareness, environmental advocates and public interest groups have relatively little access to the policymaking process due to the insular nature of party-based politics in Argentina and the weakness of the judiciary.

In both nations, federal agencies establish national programs and guidelines, but implementation is left to state governments except in cases of public lands or cross-state issues. The states or provinces in which major industry is concentrated tend to have large urban populations (Buenos Aires province in Argentina; in Brazil the states of São Paulo, Rio de Janeiro, Minas Gerais, and Rio Grande do Sul), and their environmental regulatory agencies are better funded and staffed. It is also in these states where major chemical operations are concentrated. Chemical production tends to cluster within a few areas, mostly near urban centers, where infrastructure is available to support the delivery of primary materials, mostly oil, and the storage and distribution of finished products.

By far the largest production centers remain the chemical "poles" located in São Paulo state, Bahía, and Rio Grande do Sul in Brazil, and in Bahía Blanca in Argentina. Chemical manufacturing plants are conspicuous, highly industrial, and required to operate formally under contract and license with local regulators. For these reasons, major chemical operations are relatively easy to regulate. However, even in these cases bureaucratic ineptitude and the lack of trained, professional state auditors may limit the frequency and stringency of audits and other regulatory efforts. Smaller chemical operations, or the widespread use of toxic chemicals in various small-scale manufacturing processes, pose a greater regulatory challenge because they are harder to find and often unlicensed. These facilities are far more likely than larger, more visible, and more capital-intensive plants to dump their waste untreated into the air, the groundwater, or the local sewage system.

THE CASE STUDIES

Chapter 4 examined the cases of the Forest Stewardship Council chapters in Argentina and Brazil. In those cases, factors that affect the supply of the FSC outweigh demand-side factors in determining regime effectiveness. Of these supply-side factors, the most important are a preexisting network in Brazil of forestry officials, certifiers, and environmental NGOs with interest in certification; the early participation of key business leaders; and a general attitudinal acceptance of the importance of environmental responsibility—the legacy of Brazil's special history. In addition, the influence of two sets of actors hypothesized to be critical to the success of these chapters, governments and transnational firms, proved limited. We turn now to find out if these patterns hold in the case of the Responsible Care regime in Argentina and Brazil.

The following analysis compares the size and strength of the Responsible Care national chapters in Argentina and Brazil, in accordance with the analytical model discussed in Chapter 2. This comparison focuses on the potential influence of four factors: market demand, government action, transnational firms and NGOs, and the structure of these national industries. For national chapters to be effective, demand for these privately run regulatory regimes, perceived and acted upon by firms and their associations, must be met by sufficient supply. That is, firms as a collective group must want these regimes and be capable of coordinating their actions to run them effectively.

RESPONSIBLE CARE EFFECTIVENESS

SIZE

Participation

Argentina's Responsible Care chapter was established on May 28, 1992, with the title *Cuidado Responsable del Medioambiente* (CRM), which translates to "Responsible Care of the Environment." Ninety-one company members of the Argentine Chemical and Petrochemical Industry Council (CIQyP) signed the contract. Yet this was not the first time they had an opportunity to do so. The U.S. firm DuPont had advocated local adoption of the regime to the industry council on several occasions during the preceding year and a half (1991–1992), but members had shown little interest.[8] Other American corporations who were clients and partners of DuPont also supported the idea of establishing Responsible Care in Argentina, but together these companies made up only a small portion of CIQyP members. Until late spring in 1992, these proponents had failed to convince their colleagues of the regime's usefulness.

What changed their minds was a federal order to close a petrochemical plant in a suburb of the capital, and the arrest of that company's CEO and the plant's chief manager for environmental crimes.[9] For years, local media and activists in Argentina had raised awareness of rising toxicity of local rivers and air, due in large part to industrial emissions. One of the two main rivers that embrace the capital city Buenos Aires, the Riachuelo, had been recognized for years as among the most polluted waterways in the Americas. But the arrest of two corporate managers for personal liability posed a new, more serious type of threat (although days later they were released). The headlines describing the chemical managers' arrest sent industry officials scrambling to respond. A month later, the CIQyP celebrated the creation of *Cuidado Responsable del Medioambiente*.[10] This follows a pattern, in that the origin of Responsible Care in the United States and in Canada was also aimed at countering negative publicity and preventing more stringent environmental laws and enforcement.

Participation in CRM was at its highest during the chapter's first months, but over time it dropped steadily from a high of 93 in 1992, to 75 in the late 1990s. After the financial meltdown of late 2000, membership fell to a low of 55. Currently, regime administrators and industry officials believe that participation in CRM will remain at or near its current level.[11] As of 2005, there were no plans or resources to promote the national chapter's growth.

These totals, however, represent only the companies that signed on to CRM. As Diagram 6.1 shows, the number of firms that participate actively (that is, attend chapter meetings and/or submit annual self-reports) is significantly lower. In 2000, chapter administrators offered membership to companies who provide

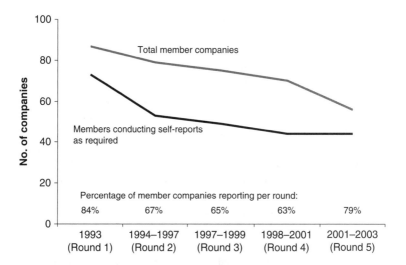

Diagram 6.1   Membership and self-reporting rates within CRM, 1993–2003

Source: Data from Cuidado Responsible del Medioambiente, interview with administrator, November 2, 2004.[12]

transport for chemicals and petrochemicals clients. Twenty-three transport companies signed on, largely because two leading transnational firms indicated they would give preference to CRM participants.

Diagram 6.1 reflects the steady decline in participation rates and the difficulty administrators have faced trying to get member firms to comply with chapter rules to self-report. This chart does not include transport companies, for which membership has been an option since 2000. In the second round of self-reporting by these transport companies, an average of 9 out of 21 companies turned in self-evaluations (43 percent). CRM coordinators and participants expressed disappointment with this low level of compliance.[13]

A list of CRM participants was not available due to a privacy agreement. Administrators stated, however, that membership in CRM includes both TNCs and local companies. The consensus view among industry officials is that American companies, first, European companies, second, and subsidiaries of U.S. and European firms tend to be the most active in CRM and are the program's chief advocates within the CIQyP. All admit that one significant shortcoming of CRM thus far is the absence of small and medium-sized firms—which is true also of the chemical council itself. Membership in the CIQyP, which requires annual dues, is widely viewed as an option for larger firms with greater revenue stability and longer-term political and regulatory needs, instead of small firms that typically operate on a quarter-by-quarter basis. Many small chemicals firms are likely to be completely unaware of the Responsible Care regime or its Argentine chapter.

European chemical companies tend to participate less, and less actively, in CRM because they view it as principally a U.S. or Canadian regime of little benefit for them.[14] Several major European transnationals do not participate at all, though in Brazil their subsidiaries do. These companies' managers report that though they admire the values and aims of the Responsible Care regime, it is not useful to them because they already have in place internal management systems that they believe are equal or superior to those of CRM member firms, and are compliant with ISO 14001. Other managers note the "incoherence" or lack of seriousness of CRM, and point out that ISO 14001 certification is recognized in international markets while Responsible Care participation is not.[15] One manager at a major German company stated:

> We have no problem with the Cuidado Responsable. We agree completely with its principles and objectives. But as it is here, it is too lax for us, as a system of certification. Our own management standards are far more strict and specific, so there's no pressure to participate and no reason to. . . . Cuidado Responsable has no teeth, no real enforcement, and everyone knows it. It is meaningless here [in Argentina], only symbolic.[16]

Since 1998, participation in Brazil's Responsible Care chapter, titled locally *Atuação Responsável* (AR), has been mandatory for all members of the Brazilian chemical industry council (ABIQUIM), which in 2005 totaled 176 firms.[17] ABIQUIM members include all of the largest chemical producers in Brazil, virtually all of the transnational firms operating in Brazil, and all firms with facilities in the major petrochemical production centers in Camaçari, Bahía; Paulinia, São Paulo; and Triunfo, Rio Grande do Sul. Among these groups there is significant overlap. As is common in highly developed chemical sectors, the industry is tightly concentrated, with two dozen firms accounting for the majority of total production. Although they make up less than 10 percent of all chemical firms in Brazil, ABIQUIM members are responsible for more than 80 percent of the nation's chemical production. Of ABIQUIM's 176 members, 43 are associate members that do not produce any chemical product, including around two dozen transportation companies and a handful of consulting and services firms that service the petrochemical industry.

Participation in Atuação Responsável, as with ABIQUIM membership in general, has been relatively stable since 1998. One recent study found that AR participants tended to be large, publicly owned, either partly or wholly foreign owned, and exporters (Roberts 1999). Making AR participation obligatory for ABIQUIM members forced more domestic, nonexporting firms to participate.

Nevertheless, the largest foreign and domestic companies still dominate the leadership of AR through their frequent participation in the chapter's 15 technical and management commissions (ABIQUIM 2005).

As is the case in Argentina, Atuação Responsável has largely failed to attract participation from more than a few dozen of the 2000 or so (estimated) small or medium-sized chemical operations.[18] Officials at ABIQUIM and participating firms identify this problem as the chapter's foremost shortcoming and a major concern. Administrators hope that a new set of program guidelines and compliance verification guides introduced in the fall of 2005, which are designed to be more flexible and accessible for smaller firms, will attract more participation from small and medium-sized firms.[19]

Another factor that influences participation is geography. AR members tend to represent disproportionately the state of São Paulo, and to a lesser extent Rio Grande do Sul. The regulatory agencies in these states are widely recognized as among the most professional and effective in the country. The common explanation is that firms in these states already had high environmental management standards, due to more strict regulation, and so for them participation is less costly or demanding.

Also, AR membership and implementation are extremely high among firms that operate within large industrial zones—of which São Paulo state has several. Within these clusters of major facilities, most particularly the enormous petrochemical centers established in the 1970s and 1980s, information and assistance is easily obtained and firms can share the costs of infrastructural improvements, services, management, and monitoring that support superior environmental practice. Also, close and cooperative operations encourage interfirm transfers of information, sharing of technology, dialogue, and the building of personal networks. Firms with operations in these industrial poles tend to be large, are often transnational corporations or have joint foreign-local ownership, and are more likely than usual to service foreign clients or markets, since these centers were designed in part to produce for export markets.

Thus, several factors contribute to this concentration of AR participation among a specific type of firm. On the other hand, small and medium-sized, local, privately owned firms are more evenly spread out across the country. These are also much less likely to operate in or near one of the major production centers. Industry officials state that in many states and regions, as in Argentina, most local firms are unfamiliar with the AR chapter and, when informed of it, consider the Responsible Care regime "something for the big companies and the multinationals, not for us," or a club to which they "have not been invited."[20]

In 2004, as Diagram 6.2 shows, 104 of 142 full ABIQUIM members (not including associate members), who are also AR participants, conducted and submitted self-reports, which is a requirement of council membership. This amounts to a compliance rate of 73 percent, in line with that of

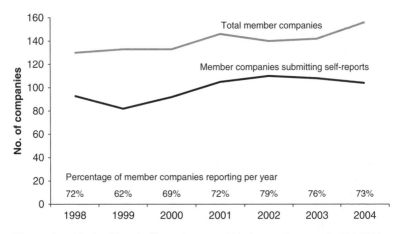

Diagram 6.2   Membership and self-reporting rates within Atuação Responsável, 1998–2004

(*Source*: ABIQUIM annual reports, 1998–2005, *Relatório de Atuação Responsável* 2005, and *www.abiquim.org.br*.)

Argentina's CRM. Around 15 percent of these members are new to AR and are given a phase-in period of two years during which self-reporting is not required. Another seven or eight members are holding companies, not required to self-report independently of their production firms. This leaves 12 to 15 "rebel" firms (as they are referred to by ABIQUIM staff) each year that do not self-report, and they receive special pressure from ABIQUIM. If within the following year these firms continue to refuse to turn in a satisfactory self-report, they are referred for action to the ABIQUIM board of directors and, in some cases, are expelled from the council.

An ABIQUIM official explained that "every year there are some. We have to call them, over and over, and then go there, knock on their office door, really bother them, all so that they do what they have promised to do. . . . Having to deal with these resistant, skeptical managers is the worst part of my job."[21] Four companies have been expelled from ABIQUIM for not complying with their obligations under AR. Two decided at that point to comply, and rejoined the council. Two remain nonmembers.[22]

STRENGTH

Impact on Behavior

Of all the managers interviewed who work for member firms of Argentina's CRM chapter, not one reported any specific change in practice or policy as a result of CRM codes or audits. One manager at a French-owned company stated that "Cuidado Responsable would be a good management tool for companies that have no management to start with. It provides a good list of practices and areas where most companies could improve their management."[23] However, many others reported having obtained ISO 14001 certification for their environmental management system, and most considered it unlikely that a firm would have no management system at all, except for small operations.

Argentina's CRM asks that members conduct audits, performed by the auditing division of the national standardization organization IRAM, every two years. Audit results yield a score between one and five. Firms that score above a four receive from CRM a certificate of environmental management excellence. Scoring below a four carries no penalty but is considered extremely rare. This is partly because the companies that agree to undergo auditing tend to be those committed to environmental management, and partly because the auditing process is designed to be flexible. Struggling to maintain the members it has, and anxious not to dissuade others from participating, CRM has little incentive to sanction any members.[24] Of the firms I visited during the interview process, none had posted a CRM certificate to be visible to public visitors, though they had hung certifications from other environmental or corporate social responsibility programs. One manager explained:

> No one knows what [CRM certification] means. Hardly anyone outside the industry knows what Cuidado Responsable is. And people inside the industry would laugh if you hung up a score report from CRM. Every member that chooses to get audited receives a score around four or five. It's like a reward you get for agreeing to be audited. Sounds ridiculous, but it's true.[25]

Industry officials reported that CRM audits were not stringent or thorough, and provided little benefit in terms of suggestions for improved management. One manager called them merely symbolic, citing the fact that audits last for half a day, regardless of the size or complexity of the unit being audited. In the case of his company, the auditors rush through. After the last two audits, the company received results certifying good management over processes that are not in place at that facility.[26] A manager at another firm stated that if few members of CRM send in self-reports, many fewer are ever audited.

> To be frank, Cuidado Responsable is a disaster. There are no real requirements for membership, so no one takes it seriously. Many members have never had an audit and never intend to. Many, including my company most of the time, fail to conduct our self-evaluations. They interfere with more important duties. Only a few people attend CRM meetings. Who cares? What purpose does it serve? I honestly can't tell you. Though I think it's a good idea, if it were seriously done.[27]

CRM has never expelled a member, nor even placed a member on probation, though the chapter's guidelines recommend taking those steps for cases of noncompliance. According to CRM administrators, some companies have been informally told that they should bring up their scores. Their response, however, was just as often to withdraw from CRM instead of improving their management.[28] CRM's chief administrator at the chemical industry council put it bluntly:

> The worst part is that our own members are not committed [to CRM]. I think they just don't care, and don't believe in it. If your own members don't believe, then how can anyone on the outside ever believe. . . . I've been running this for twelve years, and I have no illusions. My objective now is simply to keep it alive.[29]

In Brazil as in Argentina, levels of participation in Atuação Responsável differ dramatically between small and large firms. Small producers with few employees or capital and with relatively primitive operations often give little or no thought to their environmental impact, and instead seek the least expensive and difficult way of dealing with waste. It is not surprising that, among ABIQUIM members, the small and medium-sized firms report that participation in AR has had relatively major effects on their practices. Modifications reported include improved inventory and informational management, safety improvements at the plant, opening dialogue with the community, and the application of new technologies.[30] For these smaller firms, AR's guidelines and codes often provide their first map into areas of management including environmental risk, impact assessments, and community outreach, which they had never before approached in any systematic way. While the costs of participation in AR are relatively high for these low-capital firms, the few that do participate report having gained the most in terms of improved management, better safety, lowered emissions and environmental risk, and greater awareness overall. One manager at a small firm was the only one interviewed who could provide a detailed example of improved environmental performance that came from information gained via AR membership.[31]

As was the case in Argentina, few officials at large companies, foreign or Brazilian, reported specific changes in practice resulting from participation in *Atuação Responsável*. All managers interviewed from these firms, however, described the regime as beneficial: mostly for the industry as a whole, more so than for themselves. These managers point out that their practices and standards were already high before joining AR. The programs' audits and guidelines are a complement to their already satisfactory internal management systems. Many of these firms' main facilities operate within one of the country's major petrochemical production poles. By clustering their operations, these firms lower the costs of investing collectively in joint management and self-regulation, and can share the costs of creating center-wide practices and institutions that ensure high standards of safety and environmental risk management.[32]

Some managers argued that the identification of specific changes in practice is irrelevant, because the goal of the regime is to create a culture of continual improvement instead of to promote specific technologies or practices. In their view, the exchange of information, experiences, and technical solutions that occurs at regular AR meetings has improved process management and efficiency at most, if not all, member firms, simply by providing new ideas and attitudes.

Several managers reported that the effects of Atuação Responsável on their firm's behavior has, over time, diminished and shifted in its orientation. The modification of practices was most pronounced in the mid- and late 1990s, when AR promoted the new concept of integrated systems of management. This coincided with the spread of the ISO sets of standards for process quality (ISO 9000) and environmental management (ISO 14000). Since those new benchmarks for integrated management were incorporated through ISO certification and/or AR participation, few other new instruments or ideas have emerged.

These reports support the charge brought by some critics of the regime that Responsible Care only encourages reforms that are relatively easy and superficial, or that clearly add to a firm's bottom line through cost savings or added efficiency. Beyond helping firms to pluck these "low-hanging fruit," AR may have little impact. According to one Brazilian manager, "Atuação Responsável has become stagnated. Now that companies have high compliance rates and good efficiency, and talk to their neighbors and all this, they are not interested in doing more."[33]

Industry officials and ABIQUIM staff describe an evolution in awareness, and in managerial focus, from improving process safety and efficiency toward more complex, comprehensive, and externally oriented goals, such as enhancing a firm's "social responsibility." When asked where AR has had its greatest impact, the most frequently cited practice area was community dialogue. Improving or securing product stewardship is viewed as the most significant challenge AR members currently face. In the case of large firms, the AR chapter has affected most their relations with local communities, for example by encouraging the creation of Community Advisory Panels and the implementation of open-door policies at production facilities. Managers frequently referred to environmental management practices as part of the broader area of process management or risk reduction, which complicates the task of pinpointing the source of these changes.

Brazil's largest chemical firms, including TNCs, differ dramatically in the importance they impute to Atuação Responsável as a tool for their internal management. A manager at one major European firm claims it uses Responsible Care guidelines and procedures exclusively, at the global level, to manage all its safety, health, environmental, and stakeholder relations policies.[34] Officials from this same firm in Argentina, however, attributed to Responsible Care much less practical importance. For others, the guidelines and compliance verification system pertaining to Atuação Responsável serve as a reference or benchmark that complements their own internal management systems. Many managers stated that AR guidelines and codes are equal to those of the ISO's 14000 series of environmental management standards, or the U.S. OSHA's 18000 family of workplace safety standards, and so participating in AR involves at the same time meeting the terms of these other standards systems. One official explained that AR participation had prepared the firm well to certify under ISO 14001 and OSHA 18001, which, although less comprehensive, are more important since they are independently certifiable.[35]

FACTORS THAT INFLUENCE THE EFFECTIVENESS OF THE RESPONSIBLE CARE

DEMAND SIDE

Market Demand for Responsible Care in Argentina

Responsible Care's system-based standards and self-reporting do not support product labeling. Argentine member firms report marginal pressure for Responsible Care participation via client preferences, but these are difficult to isolate from demand for various management procedures. Some U.S. member firms claim to prefer, in their contracting of suppliers and distributors, those that participate in Argentina's CRM. This includes one major U.S. transnational firm that is responsible for an estimated 80 percent of all chemical production and transport in Latin America. For no company, however, is membership mandatory, and this preference in practice seems informal and flexible, especially in the case of suppliers or distributors deemed "strategic" to the interests of the company.[36]

Moreover, when pressed, most officials admit that they are less interested in CRM participation per se than they are in the fact that most CRM participants tend to be well managed, more transparent, and more responsive in all aspects of their business. Only one manager at a major U.S. company emphasized the importance of partnering with CRM participants for the reduced likelihood of his firm finding itself associated with some environmental scandal or problem.[37]

Companies that are not CRM members tend to place more emphasis on ISO 14001 certification than participation in CRM, partly because of preference within European markets for ISO certifications. Some officials mentioned programs in other industries, such as the "Q1" program from Ford Motor Company, which requires all direct suppliers to have ISO 14001 certification. Nothing of this sort is established within the chemical industry, although administrators did consider making CRM participation mandatory for all companies that transport hazardous chemicals (as have Brazilian Responsible Care administrators).

Nonmarket Demand for Responsible Care in Argentina
The threat of state regulation seems to play little role in companies' calculations regarding participation in CRM. Officials frequently describe state regulators as incompetent and/or corrupt, due to low public sector wages, the lack of technical training or knowledge of the industry, and scarce state resources.[38] Legislation aimed at controlling pollution and environmental degradation tends to be imprecise and without any technical grounding, and therefore unenforceable. Actual emissions limits are often unspecified, leaving it up to plant managers to interpret terms such as "sustainable levels," and regulatory agencies lack the technical expertise to establish those definitions.[39] As is the case with forest management certification, chemical companies seeking to ensure full legal compliance in order to obtain certification reported pressing local government agencies to clarify regulations and licensing requirements so that the companies could fulfill them.[40] This lack of effective regulation is commonly put forward as a major disincentive for participating in CRM or investing in environmental management in general, especially for smaller firms.

When asked why environmental management is such a neglected issue within the chemical industry in Argentina, interviewees both outside and inside the industry frequently cited a "backward," "short-term," or "nineteenth century" mentality among Argentine corporations.[41] Several officials opined that Argentines possess a cultural predisposition to noncompliance and evasion regarding the rule of law. Others explained this short-term thinking as a result of the experience of living through several boom and bust economic cycles, worsened by a distrust of the state. On multiple occasions subjects stated flatly, "[Argentina] is not a compliance-oriented country," or offered the local adage "*Hecha la ley, hecha la trampa* (With the law comes the loophole)."

Market Demand for Responsible Care in Brazil
As in Argentina, the fact that Responsible Care does not support product labeling means, in Brazil, that there is no end product consumer demand or price premium for products produced by AR participants. ABIQUIM officials would like to create some type of label in the future. However, Responsible Care's global directorship at the International Council of Chemical Associations is opposed to moving in that direction.[42]

Pressure from clients down the chain of production is also minimal, except in specific product lines. Most managers report that clients in other industries are unfamiliar with the AR program, limiting its usefulness as a marketing point or competitive advantage. In contrast, ISO 14001 certification is increasingly a market norm, especially in export-oriented sectors.

In some sectors, however, client-based pressure from transnational name-brand companies has provided an incentive for AR participation. In 2000–2002, major users of chemical products, particularly in the automobile and cosmetics industries, initiated formal programs to encourage their suppliers to acquire independent environmental management certification. This pressure from a critical set of clients jolted ABIQUIM. The Ford Motor Company, for example, told the Brazilian division of a major German chemical firm that compliance with AR was insufficient to meet the standards of its new quality control program. Ford officials told the chemical producer that it must pursue independent certification, such as through ISO 14001, or its contracts with Ford would be at risk.[43] This firm, in response, led a campaign within ABIQUIM to revise the AR program to include mandatory independent audits and certification, closer to the ISO 14000 model.[44] The result was the *VerificAR* system, introduced in 2005, which requires biannual audits by a team of independent observers including community representatives and certified independent auditors.

However, even direct pressure from automobile producers, and more recently from a major global cosmetics company, did not lead to any observable change in AR participation rates. Most chemical firms serving these manufacturers were already members of AR, and furthermore already possessed ISO 14001 and other certifications. The firm caught off-guard by Ford Motor Company's change in requirements was the firm that had committed to making Responsible Care its management verification system of choice throughout its global operations. The key point is that for most firms already plugged into global production chains through clients like Ford Motor Company, achieving independent environmental management certification is relatively easy and affordable. Likewise, *VerificAR* poses little difference to these firms. Officials familiar with the 14 *VerificAR* audits that had been performed by the winter of 2005 reported few surprising results and no significant change in practices as the result of the new system.[45]

ABIQUIM's principal goals in creating the *VerificAR* system were to cut redundancy across standards and audits, simplify AR to make it more accessible for smaller firms, and incorporate independent certification.[46] *VerificAR* incorporates within a single audit AR verification, ISO 14001 certification, and scoring for the National Prize for Quality (*Prêmio Nacional de Qualidade*), a popular, state-sponsored annual competition awarding the best-managed firms in Brazil. Independent certification will likely produce more stringent and detailed evaluations of compliance with AR guidelines, including in the area of environmental management.

Nevertheless, to skeptics even the *VerificAR* system remains too flexible in its standards and metrics, is insufficiently transparent, and bends too much to firms' preferences to be accepted as a credible regulatory tool. For example, ABIQUIM still prohibits allowing public access to certification data or reports, and the identities of auditors, audit dates, and periods of validation are all withheld, though Ford's Q1 program and ISO 140001 certification require these to be made public.[47] In terms of the impact of *VerificAR* on AR participants, it seems most likely that large firms seeking multiple certifications and awards will benefit from the reduced frequency of audits and the multipurpose nature of the one annual audit. Small or medium-sized companies, on the other hand, must submit themselves to more comprehensive, and more costly, audits. ABIQUIM officials expect, however, that the effect on small companies, especially new AR participants, will be minimal due to generous phase-in provisions and the additional flexibility that *VerificAR* allows firms in defining their own targets for performance.

As in Argentina, the promotion of the AR national chapter by major transnational firms seems mostly rhetorical. Only one chemical TNC claims to have a formal program to advocate AR participation among its suppliers and contractors, and it falls short of making participation a contract requirement. Other AR participants, both foreign and domestic, claim to prefer convincing other firms of the benefits of participation in terms of improved, more integrated management systems, instead of pressuring them to join.[48]

ABIQUIM has recently created a safety and environmental management certification program designed after AR but applicable to transport services providers, and has made it requisite for contracts with ABIQUIM members. This move was controversial even within ABIQUIM because it created barriers against many service providers. Several officials viewed it as favoring a handful of large, specialized transport firms to the disadvantage of smaller, local transport companies that serve a range of clients beyond chemical manufacturers.[49]

Nonmarket Factors that Influence the Effectiveness
of Responsible Care in Brazil
Managers do not report pressure from regulators or from health or environmental advocacy groups as significantly affecting their decisions to participate in AR. The most significant recent piece of relevant environmental legislation, the Environmental Crimes bill of 1998, did not make any noticeable difference in participation rates, though it is widely held to have caught the attention of industry.

Governments can influence corporate participation in private regimes by offering incentives and imposing disincentives. Chemicals company managers and AR administrators say that, in regard to AR in regard to participation the

Brazilian government does neither effectively. Only one firm surveyed reported that AR participation had any effect on its relations with local regulators. The information required by AR self-reports regarding inventory, emissions, waste, and the handling of toxic materials facilitates, for this small firm, its audits for state agencies. Large firms report no effect in terms of their relations with regulators. The only exception are the cases of firms with operations at the Camaçari petrochemical pole in Bahía state, where AR participation is mandatory and organized collectively through an independent management firm, which also handles relations with regulators.[50]

For their part, Brazilian regulators are skeptical that AR participation translates into full legal compliance or best practices. Nevertheless, they agree that many of the firms that participate in AR—the largest chemical producers, especially TNCs—tend to be legally compliant and require little special regulatory attention. These large firms typically operate in concentrated production zones where tight regulatory control and full legal compliance have been the norm for 20 years, long before AR was created.[51] It is not these large firms that cause regulators to worry, but smaller, low-capital, loosely managed producers who lack information and resources to concern themselves with environmental impact. As one industry council staff person explained, "ABIQUIM members produce eighty or ninety percent of the chemical product in the country. But it's the other ten or twenty percent that causes ninety percent of the negative environmental impact."

Similarly, industry officials in Brazil deny that pressures from NGOs or community activists have had any immediate influence over decisions whether or not to participate in Atuação Responsável. The only environmental NGO mentioned during interviews is Greenpeace, which is well known to inspect chemical facilities and their surrounding areas and has, on several occasions, made public allegations of toxic pollution. Managers could not, however, recall any particular instance of targeted pressure by the group that led to any verifiable claims of malfeasance or any wider speculation that AR would improve relations with that organization. Most managers believe that, regardless of what a firm does or whether or not it participates in regimes like AR, environmental NGOs will always scrutinize its operations and seek opportunities to make public denouncements. AR alone, therefore, provides little or no protection, beyond whatever improvements in practice it creates.

In contrast, community organizations or neighbors seem to be a relatively important source of pressure and concern. Most managers described at least one instance in which neighbors' complaints regarding odor, loud noise, smoke or steam, or liquid runoff led directly to changes in technology or practices. For this reason, the AR chapter's guidelines, best practice cases, and information regarding community and stakeholder relations are viewed as some of its top benefits. One manager explained:

> We are engineers. Give us a mechanical or technical problem and we can solve anything. But dealing with people, it's beyond us. We have no training. . . . [Atuação Responsável] presents it all in guide books with instructions, step by step, and lots of information. This has been incredibly useful.[52]

Although few individual firms feel significant direct pressure from community groups or NGOs, the industry council ABIQUIM keeps track of public perceptions regarding the chemical industry. In 2002, ABIQUIM sponsored a survey to measure the impact of AR after ten years of operations. The survey showed that Brazilians still viewed chemicals as among the most dangerous, dirtiest, and least trustworthy of industries. In response, ABIQUIM emphasized within AR the importance of community outreach and dialogue. For that purpose, in 2004 ABIQUIM created a National Community Council within Atuação Responsável that consists of independent public representatives who have the authority to review and publicly critique AR's policies and performance.

SUPPLY SIDE

Lack of Impact from Transnational Firms and NGOs in Argentina

Argentina managers commonly blame the industry council for Cuidado Responsable's lack of effectiveness. The council allocates few resources to the chapter and has refused several requests to hire full-time staff to operate CRM. The CRM chapter's total annual cost is estimated at less than US$10,000, composed mostly of part-time pay for its one administrator, the lease of meeting space, and a small travel budget. CIQyP has also refused to mandate participation in CRM, as the Brazilian and U.S. councils did in the late 1990s. The purported position of the CIQyP leadership is that CRM, CRM is a peripheral activity, strictly voluntary, and it must not pose any additional costs or difficulties to industry council members who already suffer from a difficult economic environment.

This lack of commitment to the Responsible Care regime is partly due to the dominant position held by oil and gas companies on the council's executive board. One major company in particular, a formerly state-owned but now Spanish-Argentine oil and gas firm, has significant influence over the activities of the council, but few real ties to the chemical manufacturing industry. Oil and gas companies do not typically associate themselves with the chemicals manufacturing industry, and none of the world's major oil companies participates in Responsible Care.

The regime's key supporters, as well as a disproportionate portion of its active participants, have been foreign—especially U.S.—firms. DuPont and its partner firms played a key role by introducing CRM to the

Argentine industry. Three companies interviewed began to engage in environmental management only after being purchased by a European or U.S. company.[53] This support, however, has not translated into mandatory CRM participation for suppliers, additional resources for the chapter, or any more active type of support than simply participating. One official at a major U.S. company allowed that council members are also to blame for CRM's weakness in Argentina. In his view it is unrealistic to leave the job of promoting *Cuidado Responsable del Medioambiente* entirely to the coordinators.

In Brazil, AR Success Driven by Local Industry Council

In the Brazilian case, on the other hand, a full-time staff of six chemical engineers and/or former auditors administers Atuação Responsável from ABIQUIM's office in São Paulo. The chapter runs on an annual budget of approximately US$500,000, though these costs are not differentiated from ABIQUIM's total operating expenses and are thus difficult to identify accurately. This is because, as one ABIQUIM official explained, "ABIQUIM's business is, in large part, *Atuação Responsável*. It's not ABIQUIM managing *Atuação Responsável*. *Atuação Responsável* is an essential part of what ABIQUIM does, for its members."

The staff conducts regular meetings (monthly or quarterly) of 15 different committees that review and implement chapter policies.[54] Committees are composed of representatives from member firms. AR staff also produces the annual national conference and coordinates the writing of AR codes and guidelines. The first national practice code was drawn up in 1996. Since that time AR has written several codes of practice with technical guidebooks on implementation, has revised its verification system (*VerificAR*), and most recently has dramatically overhauled the local program to make it more accessible and its codes more applicable to the council's diverse members.

As with Argentina's Cuidado Responsable, the original principles and codes of practice were derived from those of the U.S. Chemical Manufacturers' Association. ABIQUIM, however, elaborated standards, codes, and practices distinct from those of the United States, including most recently the *VerificAR* system. These modifications have come in response to member firm demands (e.g., one major firm's promotion of independent certification) and challenges (e.g., simplifying the program to increase accessibility for smaller companies).

Brazil's AR chapter does not at the national level collaborate with, or even have contact with, any NGOs, community groups, unions, or government agencies. One of the chapter's codes of practice is the establishment of community councils at the firm or facility level, through which community representatives can learn about and express their views on local practices.

Several firms and researchers identify these councils as an important and useful step toward an enhanced public image and improved corporate responsibility, especially in the case of the chemical production center in Camaçari, Bahía.[55]

As mentioned above, in 2004 AR created a forum for public representation at the national program level. This National Community Council consists of a diverse group of notable individuals, including popular singers, university professors, and community leaders, nominated by member companies. This Council meets twice a year and reviews AR program activities and policies. Member company officials describe this as a progressive first step in the improvement of public awareness of the benefits of Atuação Responsável. Skeptics both inside and outside the industry, however, question the usefulness of a council whose membership and scope of operations are steered entirely by the executive council of ABIQUIM.[56] AR administrators admit that the purpose of the National Community Council is to improve public awareness of the existence of AR and raise its national profile, rather than to engage these individuals in a serious review of the chapter and its methods.[57]

The Irrelevance of State Actors in Argentina

Argentinian state agencies play no supportive role nor provide companies any incentive to participate in CRM. The two existing federal programs that aim to promote enhanced corporate environmental management and accountability have had no contact with the chemical industry's CRM chapter.[58] Coordinators at the state agency and CRM administrators both agree that such cooperation would be useful. However, both lack the resources and support from their directing institutions to attempt a public-private partnership.[59] As the administrator of the federal "Clean Production" initiative stated:

> Here in the government, the common thing is for new programs to be created, and old programs scrapped, by each new administration or agency chief. They are often seen as personal items, belonging to the former chief or the former group in power, not permanent programs. So it seems every two or four years or so everything gets turned over, new faces are put in everywhere, and we here on the ground start over from scratch.[60]

The Role of State Actors in Brazil

Atuação Responsável has no collaborative relationship with any Brazilian government agency at the national level, nor do regulatory officials or ABIQUIM administrators have any initiative to explore such ties. Regulators do not consider AR's standards or verification system relevant to compliance with local or national laws. The only exception to this lack of interaction between AR and national government is the representation of AR, through ABIQUIM, on a federal council that oversees national law on chemical safety and emergency preparedness. The usefulness of the Responsible Care regime as an industry initiative is noted in the council's official reports.

Collaboration between AR and regulatory officials is infrequent also at the state level. The only sustained case of interaction regards environmental control policies at the enormous industrial center in Camaçari, Bahía. The Bahían state environmental regulatory agency offers firms operating at Camaçari a more collaborative, compliance-oriented style of regulation. This more transparent and cooperative relationship is facilitated by a local environmental and quality management firm, COPIC, which is paid by the firms at the industrial center to coordinate and supervise the safety and environmental management practices at the complex.[61] COPIC also encourages and facilitates AR membership, and has argued for local regulators to include AR reports as part of the audit process.

According to chemical managers at facilities in Bahía, AR members across the state receive special treatment in the licensing process, and a public-private program encourages AR participation among these companies' suppliers. This Bahían case is gaining national recognition as a promising model for further antipollution policies. Regulatory agencies in São Paulo and elsewhere are considering implementing aspects of these programs into their own operations, most likely starting at the industrial centers where regulation is especially stringent.

Many of the supply-side factors that influence the effectiveness of the Responsible Care chapter in Brazil are contextual and their influence indirect. First, the size and resources of the chemical industry in Brazil dwarf those of all other chemical sectors in South America. This gives the industry and particularly its member-driven organization, ABIQUIM, more resources, both financial and human, to coordinate and administer a national-level chapter. The fact that in Brazil dozens of large, diversified, competitive domestic chemical firms survived and grew during liberalization also helps account for the industry's ability to support AR. Firms with larger scales of operation obtain greater returns from their investment in new technologies or management systems. These efficiency gains increase if the firm can share some of those costs across all partners at an industrial center, as is done at the Camaçari complex. The relatively high number of large chemical firms in Brazil, compared against Argentina, and the clustering that increases shared costs and spillover effects from investment in environmental management, increases incentives for companies to participate in programs like AR.

Another important factor is the historical openness and capacity for collaboration that Brazilian industries have shown toward foreign capital, a legacy of the state-led "triple alliance" projects of the 1970s and 1980s. Brazilian managers, and independent certifiers who work across South America, describe Brazil's business culture as relatively open to foreign models and practices, and more risk acceptant, than the more conservative and rent-seeking attitudes often found in Argentina or Chile.[62] Instead of suspicion and trepidation, and a dependence on the state to keep foreign capital at bay, this attitude on the part of many industrialists in Brazil has created a business culture where innovative institutions such as AR can thrive.

In addition to being open to foreign practices and models, Brazilian industries have a long history of national philanthropy, justified either as corporate responsibility or as an element of Catholic service. Long before Responsible Care, Brazil's largest industrial companies and banks created numerous national programs in the areas of business ethics and social accounting, some of which have lasted for decades.[63] National Catholic business associations and other organizations promoting philanthropy and social responsibility have existed in Brazil for almost a century. Moreover, due to public concern over the loss of the country's rainforests (both in Amazônia and along the Atlantic coast), Brazilian society has for decades been sensitized to environmental issues. These national experiences with environmental crisis, in regard to both forests and toxic pollution from chemicals manufacturing in Cubatão, São Paulo, lay the attitudinal groundwork for the later success of contemporary private environmental regimes.

EXPLAINING THE GAP IN EFFECTIVENESS

As with the cases of the Forest Stewardship Council, the Responsible Care chapter in Brazil is superior in scope and strength to its Argentine counterpart. Though both chapters cover the same practice areas, as defined by the ICCA at the global level, Brazil's industry council offers firms a much more detailed and comprehensive set of guidelines. In Argentina, CRM's administrator offers regular meetings where managers can discuss environmental issues and the implementation of CRM guidelines and standards. In contrast, the Brazilian chapter regularly convenes 20 different groups to discuss the implementation of Responsible Care within specific technical areas. Moreover, to make the regime more useful to its diverse membership, the Brazilian chapter includes guidance councils for ten different sectors of the industry. By differentiating among its types of member firms, and refining the program's codes and guidelines to address their individual needs, RC administrators in Brazil have expanded the regime's scope over time to enhance the benefits they can provide their members.

Brazil's chapter is also impressive in terms of the strength of its demands and sanctioning power. Brazil's chapter has made participation in Responsible Care mandatory for all chemical council members. Administrators have also expelled firms from the council upon grounds that they were not complying fully with the program's reporting and auditing requirements. These policies, rare in developing countries, mirror those in place in Responsible Care's home countries, the United States and Canada. Argentina's chapter is more typical, in that participation is voluntary and the chemical council is more concerned with maintaining its dues-paying membership than with imposing Responsible Care requirements. Also, though compliance with self-reporting requirements is relatively low, no company fears sanction from chapter administrators, who are marginalized even within the council.

Surprisingly, despite these differences in scope and strength, actual participation levels in terms of meeting self-reporting requirements are similar, at around 70 percent. This indicates that although a greater share of Brazilian chemical firms participate, of those that participate a similar share tend to shirk their reporting and auditing demands. The problem that Andy King and Michael Lennox (2000) identify in regard to the U.S. Responsible Care chapter applies in these cases as well. Without sanctioning power, Responsible Care is unable to control the free ridership of firms that sign on to the regime but do little or nothing to implement changes in line with its objectives.

Unlike the FSC cases, the Responsible Care chapters demonstrate similar levels of formal participation despite dramatic differences in the quality and services of the two national programs. In both nations most of the industry leaders participate, and participation rates are particularly high among transnational firms, though in Argentina European firms and their subsidiaries seem generally less inclined to participate than U.S. firms.

While it is difficult to gauge with precision, Brazilian firms report more changes in their practices as a result of Responsible Care participation than do their Argentine counterparts. In the view of most Argentine managers, Responsible Care does not particularly define or encourage superior environmental practices as much as it endorses the various enhancements that leading firms make of their own accord. In Brazil, managers describe a more applicable and technically useful set of program guidelines and frequently mention several examples of practice areas affected by participation. In neither country, however, does this regime's effects on members' practices compare with those reported in the FSC cases, where standards are performance based and independent verification is in place.

In regard to the factors hypothesized to have influenced the effectiveness of these national factors, these cases offer further support for the findings from the FSC cases. In these RC cases, supply-side factors and conditions have had significantly more impact than demand-side factors on the effectiveness of national programs. The common depiction of market returns and supply chain pressures as the engine of private environmental regulation does not apply in these cases. Without product labeling, Responsible Care does not provide broad market benefits. However, as research in other nations has shown about other nonlabeling regimes such as ISO 14001 standards, we should expect supply chain pressures driven by transnational firms based in northern markets to be in evidence, but they are not. Despite managers' claims that large foreign and domestic firms give preference to regime participants in their contracting, there is no evidence this is so.

Evidence from these cases also fails to support two other assertions common in research on private forms of environmental regulation. In these national cases, state regulators have little or no direct impact on Responsible Care's effectiveness. State actors have not endorsed the regime, nor do they give any preference to regime participants in their audits, requirements, or (as purchasers) in their contracting. With the exception of one industrial center in the northeast of Brazil, firms do not report changes in their relations with regulators as considerations affecting whether or not to participate, nor to what degree.

Nevertheless, as with the cases in the forestry and wood products industries, state actors influence these regime chapters because they are major elements of the national regulatory and business environments in which these regimes must operate. In Brazil, for example, the relatively high degree of professionalism of many environmental control agencies—at least in industrialized, urban regions—has over the years helped to shape the culture of its chemicals industry. Because for two decades Brazil's leading chemical firms had faced an effective environmental control regime and intense public scrutiny, they were already interested in environmental management and risk reduction when Responsible Care was introduced in 1991–1992. The purported benefits of the regime made immediate sense to these Brazilian firms, whereas in Argentina local firms showed little interest in the regime until a federal judge shocked the industry by throwing chemical facilities' managers in jail.

The international diffusion of Responsible Care has been portrayed elsewhere as the work largely of major U.S. and European chemical firms, who use their market share and hegemonic influence within industry associations to promote the regime's acceptance abroad.[64] These case studies, like those of the FSC, suggest that this account is accurate only for the earliest stages of the regime's diffusion. Further on, however, as the excitement of the initial announcement of the regime fades and implementation demands actual investment and participation on the part of firms, the significance of advocacy by major transnational firms decreases. Local firms and leading personalities within the industry either step forward to administer and encourage the national chapter, as in Brazil, or they remain passive as in Argentina. Also, the influence of transnational firms is limited to promoting formal participation. Without the support of local firms and the local industry council, these firms have no sway over the establishment or exercise of sanctions against noncompliant members.

As the summary Charts 6.1 and 6.2 demonstrate, these cases further support the argument that explanations for the variation in effectiveness of these regimes should focus less on factors or actors external to the industry, and more on the attitudes and decisions of local firms and the local organizations through which they associate. The most important factors affecting the variation between Responsible Care chapters in Argentina and Brazil concern the preexisting attitudes of local firms toward environmental management, and the organizational capacities and cultures of the national industry associations.

Chart 6.1   Summary of the Responsible Care - Argentina case

Demand-side factors *that influence effectiveness*	Negligible market benefit, domestic or international, for participation in CRM. Program does not support product labeling. Some large firms state a preference for CRM participants in contract bids, but this is not a high priority.
	Nonmarket pressures on firms from state regulators or civil society are negligible. Regulation is generally lax, despite occasional high-publicity prosecutions following particularly egregious offenses. Environmental control regime is fragmented and has little political or material capital. NGOs and community groups are not a significant source of pressure on firms.
Supply-side factors *that influence effectiveness*	Administrative agency is poorly funded and staffed. The national industry council, which includes massive petroleum companies as well as chemical firms, offers only limited support to CRM. CRM activities are limited to a single monthly meeting, sparsely attended, and annual reporting of the results of self-audits. Participation in CRM is not mandatory for council members, and CRM has no sanctioning power for participants not in compliance.
	There is no cooperation or contact, formal or informal, between CRM and any NGO or community group. Civil society groups are typically unaware of CRM, and those who know of the program view it as greenwashing.
	State regulatory or environmental planning agencies are disinterested in collaboration with or recognition of CRM, and act institutionally only on a short-term basis due to political instability, which limits their capacity for longer-term policymaking.

Chart 6.2   Summary of the Responsible Care - Brazil case

Demand-side factors that influence effectiveness	There is no consumer market demand or benefit from participation.

Demand-side factors that influence effectiveness

There is no consumer market demand or benefit from participation.

Pressures from clients exist and are increasing, especially from transnational firms in specific industries such as automobiles and cosmetics. However, these clients demand environmental management, not Responsible Care participation per se, and tend to prefer certifiable standards such as ISO's 14000 series. Small companies report program participation may yield a slight competitive advantage as indicative of technically sound management.

Threats of regulatory action are not a significant source of demand, although risk control is considered extremely important. Major firms and facilities face effective regulation and already operate above legal compliance. Smaller companies tend to face less concentrated or regular regulatory attention.

Environmental NGOs are not a significant direct source of pressure, though societal concern over environmental degradation increases managers' acceptance of environmental management as a necessary investment. Relations with neighbors are an area of higher direct concern but do not significantly influence firms' decisions regarding Atuação Responsável.

Supply-side factors that influence effectiveness

Administrative agency is well staffed with technically trained professionals and enjoys strong support from industry council and all leading firms, foreign and domestic.

Legacy of foreign and domestic joint ventures, a relatively flexible and open business culture, and an abundance of local resources' organizational capacity reduced the program's stigma as a "foreign" program.

With the exception of the local regulators at one major petrochemical center in the state of Bahía, state regulatory agencies neither endorse nor encourage participation.

Atuação Responsável has no ties to, collaboration with, or participation from organized community groups or NGOs. In creating a National Advisory Council consisting of public representatives, AR has made a small step toward incorporating nonindustry, civil society into program oversight. However, program members coordinate their own public outreach.

GLOBALLY SOWN, LOCALLY GROWN: HOW LOCAL ORGANIZATIONAL CAPACITY LIMITS THE VIABILITY OF GLOBAL PRIVATE REGIMES

This book asks *to what degree, and under what conditions, are global private environmental regimes effective in developing nations?* To answer this question, we have examined reasons why the Forest Stewardship Council and the Responsible Care regime have fared so much better in Brazil than in Argentina. Our analysis focused on four factors commonly believed to be critical to regime development and success: market demand for certified goods, transnational actors, governmental support, and industry concentration. While each case presented its own idiosyncrasies, all together these four cases suggest that none of these factors has played a significant role in determining why both private regimes are so much more effective in Brazil than in Argentina. Instead, these cases suggest that local organizational capacity—the social and material resources of local interest groups and coalitions who support these regimes—is the most important determinant of local chapter success. International demand signals, pressures across supply chains, and the activism of transnational industry groups and NGOs may be responsible for the creation of these regimes and their spread to developing nations. Once introduced, however, their effectiveness within national industries depends on the capacities and attitudes of local advocates.

This conclusion considers how this finding in regard to organizational capacity can be extended to other developing nations and explores this possibility through brief examinations of these regimes in three other South American nations. We then return to the question posed in Chapter 1 about the viability and effectiveness of these global private regimes as instruments of global governance. What do these cases of private environmental regimes in Argentina and Brazil tell us about the effectiveness of private regimes at the global level?

EXPLAINING LOCAL ORGANIZATIONAL CAPACITY IN ARGENTINA AND BRAZIL

These case study Chapters 4 and 6 explain in detail how local organizational capacity in Brazil supported an active network of environmentalists and industry leaders ready to support the FSC, and how a forward-thinking, innovative chemicals industry council assumed RC as a key competence. Likewise, they detail how in Argentina it was the lack of local organizational capacity that undermined regime effectiveness, whether it was regional and intersector splits that divided the forestry and wood products sector, or an indifferent chemicals industry council.

This study argues that, at the time of the introduction of these two global regimes into these countries' wood products and chemicals manufacturing industry, a specific level and type of local organizational capacity was in place. Furthermore, this capacity is largely immutable, despite the efforts of industry councils, NGOs, or governments to improve it. To understand why this is so, we need to consider more closely the causes of these local organizational capacities.

Diagram 7.1 captures the most relevant historical factors in each country case and highlights their effects (years or decades later) upon the structures and cultures of these industries. Brazil's chemicals manufacturing industry was more receptive to Responsible Care because of its previous exposure to public scrutiny, coupled with industrial policies that, over years, had helped the industry grow diverse and competitive enough to survive the influx of foreign investment in the 1990s. Argentina's industries suffer, in terms of their organizational capacities for supporting effective local regime chapters, from weakness due to recurrent economic shocks combined with their relative insensitivity to the need for environmentally responsible management. Viewed from this angle, historical environmental and economic conditions constitute a set of deep, principal causes for observed organizational capacity. The more immediate factors such as industry structure and management culture are secondary, though more proximate, determinants of effectiveness.

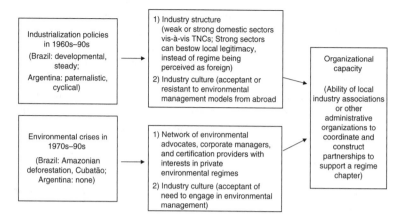

Diagram 7.1   How national legacies generate local organizational capacity

COMPARING ARGENTINA AND BRAZIL TO CASES
IN OTHER DEVELOPING COUNTRIES

Even if the importance of local organizational capacity is clear in these cases, how confident can we be that these cases are representative of most developing countries, and therefore that local organizational capacity is a critical factor for regime effectiveness around the world?

In many ways the economies and policies of Argentina and Brazil are typical of large-sized, middle-income, industrialized democracies. Though Brazil's economy and industrial sector dwarf those of Argentina, both nations feature large, relatively modern industries that grew for decades by serving significant domestic markets. Both countries also feature productive primary sectors—particularly agriculture, fishing and ranching, mining, and forestry—that have long been principal sources of economic growth. The Argentine and Brazilian governments have traditionally welcomed foreign investment and firms. Since the mid-1980s in particular, both nations have encouraged foreign investment and have sought to improve the competitiveness of their industrial and agricultural sectors via the import of foreign technologies and models.

Within Latin America these countries are relatively large, and more industrialized than most except Mexico. But their economic paths and policies are similar to those across the region, except for the sole nondemocracy (Cuba), and the poorest and smallest of nations—Haiti, Nicaragua, and Honduras, for example—which depend on foreign aid as much as on their own production. Foreign investment, transnational actors, and the state play similar roles in all these countries, and exports to the United States, Europe, and East Asia are major sources of revenue. Looking beyond Latin America, most middle-income, developing democracies including the prominent states of East and South Asia and Eastern Europe share most of these features as well. The greatest differences lie in the history and makeup of their political . institutions, not in their industrial or economic policies. At least at a basic level, there is good reason to accept Argentina and Brazil as generally representative of these countries, and patterns of events identified in Argentina and Brazil as suggestive of tendencies within their national industries as well.

Less-developed nations in which industrial sectors are small or rudimentary and/or where civil, political, or economic freedoms are curtailed, are not included in this group. The actors and conditions that pertain to these four cases in South America either do not exist or operate differently in their circumstances, and we cannot infer any similarities in terms of regime effectiveness. Where firms, NGOs, and business associations cannot operate freely, or where foreign investment and trade are minimal, there is no supply-side or demand-side basis for private regulation. Indeed, private regulatory regimes rarely operate in such countries.

In order to test the degree to which our thesis of local organizational capacity helps explain patterns in regime effectiveness in other developing countries, we now examine three cases of private regime chapters in other South American nations: the Forest Stewardship Council in Bolivia, Responsible Care in Ecuador, and Responsible Care in Chile.

The similarities across these countries' industries, economies, and political systems should not be overstated. Argentina and Brazil are more industrialized and have larger domestic markets, so that exports are a less important source of national income. Bolivia is one of the poorest countries in Latin America, and both it and Ecuador lag behind the other three (Argentina, Brazil, and Chile) in income per capita and industrialization. Chile's economy and its industrial assets are a fraction of those of Argentina or Brazil, but it has many advanced, competitive industries, particularly in agriculture, fishing, mining, and other natural resource sectors. Together, these three additional minicases present a wide range of economic, political, and industrial conditions, which is helpful for our task of exploring when and how local organizational capacity is an important factor for the effectiveness of private environmental regimes.

THE FOREST STEWARDSHIP COUNCIL IN BOLIVIA

FSC's international administrators and advocates frequently cite Bolivia's as one of its most successful national cases. Bolivia was, with Brazil, among the first South American nations to establish its own national FSC forestry standard, in 1999. As of late 2008, 20 Bolivian forests were certified compliant with FSC standards, and 24 manufacturers and handlers had obtained chain-of-custody certifications. Bolivia currently leads the world in its area of certified tropical forest, and its percentage of total forest area that is certified—almost 5 percent of total managed forests—is the highest in Latin America.[1] Observers forecast that this percentage could someday reach 10 percent.[2]

The reasons for the relatively high level of participation are twofold, and contrast with the findings from Argentina and Brazil. First, forestry accounts for 11 percent of Bolivia's foreign exports, and an estimated 50 percent of the industry's production is export oriented, a much higher percentage than in either Argentina or Brazil. As in those countries, the purported market benefits of FSC certification in Bolivia, in terms of securing access to the markets of North America and Europe, have been an important source of incentive for producers to certify.[3]

The Bolivian case also differs from the pattern in Argentina and Brazil because its government has profoundly influenced the decisions of producers to participate in the FSC. Support for a national forestry certification program has existed since the early 1990s among NGOs, community groups, and a handful of firms. However, participation surged after 1996, when a new Bolivian Forestry Law changed the fee structure and usage rights on public lands (all of Bolivia's native forests are publicly owned) and mandated compliance with norms of forest management. The new law broke up the holdings rights of large timber companies and opened millions of hectares to new activities by indigenous people, local communities, and private landowners, on the condition that they meet forest management standards. Under the new law, fees for usage of the forest are based on area instead of the volume of harvested wood. This made the holding of large tracts of forest and the harvesting of individual, select species very expensive, and many firms sold their rights instead of operating under the new regime. Also, by making usage rights more competitive, the law encouraged interest in new species and markets.

Although the government did not directly require independent forest management certification of all operators, the new law promoted interest in the FSC by putting in place similar standards of forest management, which made FSC certification a relatively easy and inexpensive additional step. It also promoted the diversification of product lines and the modernization of the industry, which encouraged exports and improved competitiveness. FSC certification grew largely because it offered a fairly easy next step toward greater competitiveness in foreign markets, along a path already established by state law and policies.[4]

There is a feeling among many within the FSC that the Bolivian case, in fact, represents too much of a good thing in terms of government endorsement of the program. In Bolivia, because certification under FSC is virtually required to secure a license to log, extract, or manage forest on public lands, regulators have come to rely on FSC certification as a regulatory tool. This bolsters demand for the regime among firms and other producers. However, it also places excessive strain on the certification providers since their work has come to involve, or in fact *to be*, bureaucratic red tape. Moreover, this delegation of government oversight to FSC certifiers introduces additional incentives for corruption, a possibility that poses a grave threat to FSC's international reputation. In terms of the two dimensions of regime effectiveness, state policies in this case have changed participation rates. However, boosting demand in this way (i.e., by aligning FSC certification with legal requirements) may reduce the regime's impact on participants' practices if political pressure is exerted on the industry's behalf or if the certification process falls victim to venality.

There are many similarities and differences between Bolivia's FSC case and those of Brazil and Argentina. Market demand is reported to be an important motive for participation. However, as in Argentina and Brazil, market benefits have not yet been realized to any significant degree. Because there are fewer domestic resources to support sustainable forest management, transnational NGOs in particular (though not transnational firms) have been much more important as the principal source of financing and support over several years. The most striking difference, however, is the strong role played by the Bolivian government. The government boosted FSC by aligning forestry laws in a way to support forest management certification as both a licensing requirement and a means to enhanced competitiveness.

The Bolivian FSC case provides further support for the claim that local, supply-side factors significantly influence regime effectiveness. However, contrary to the other cases we have seen, the government's use of the regime as a tool for the management of public forests is the central cause of almost complete participation. Government policies create demand-side pressures in terms of making certification obligatory for forest producers and boost supply-side conditions by bestowing public credibility on the regime. Other significant differences between Bolivia's case and those in Brazil and Argentina are the importance of the roles played by transnational NGOs in promoting and supporting the FSC in Bolivia, along with the absence of transnational firms as important actors.

RESPONSIBLE CARE IN ECUADOR

With 42 members, *Responsabilidad Integral*, Ecuador's Responsible Care chapter, is much smaller than its counterparts in Argentina and Brazil. As of 2005, the Ecuadorian chapter featured only three codes of practice and had poor participation levels in terms of self-reporting and verification.

For example, in 2004 only 13 of 37 members presented annual self-reports, and only two underwent the verification process. Only 8 of 37 showed indications of attaining satisfactory levels of performance.[5]

The chapter, however, has developed some promising initiatives. In 2005, Responsabilidad Integral initiated an annual contest that rewards the local companies that most improve their environmental and safety practices in line with the regime's principles and objectives, and recognizes all companies that perform well on their self-evaluations. These prizes give additional incentive to firms to participate in the program, as a means of securing public recognition of their positive performance and commitment to environmental management. Moreover, government and labor representatives take part as observers and sponsors of the competition, and actually give the awards. In the view of the director of *Responsabilidad Integral*, this enhances the credibility of the competition within the larger community. *Responsabilidad Integral* has also created a guidelines system by which certain participating companies who meet strict standards of reporting and management assurance can display the national chapter's logo on their products and advertisements.

This case is remarkable because of the support that *Responsabilidad Integral* received, from its inception until 2006, from the country's largest environmental NGO, the *Fundación Natura*. In 1998 this environmental organization approached APROQUE, Ecuador's chemicals industry association, with a proposal to establish a local Responsible Care chapter.[6] With a small chemical industry and relatively few members, APROQUE's resources were limited and it had no experience at administering this type of regime. *Fundación Natura*, a local organization but one with integral ties to the World Wildlife Fund and the Swiss Agency for Development and Cooperation, offered to fund the national chapter for a period of five years and to give technical assistance on its environmental management aspects. This arrangement was in place until January 2006, when APROQUE assumed full administrative control over the chapter.[7]

In this national case, local organizational capacity is the only factor that explains the regime's local existence and development. One local environmental NGO—albeit one with close ties to transnational NGOs—supplied this organizational capacity at the early stages and promoted the creation of capacity within the local industry association. Without this initiative, APROQUE would not have been able to implement a national chapter. Considering the typical attitudes of chemicals industry councils, including those in Argentina and Brazil, what is remarkable is that APROQUE was willing to accept the guidance of an environmental NGO in establishing this regime and its guidelines.

Considering the typical adversarial relations between environmental groups and chemical manufacturers, it is also extraordinary that the Fundación Natura sought out this partnership. This Ecuadorian model may prove to be a guide for advocacy communities and industry councils in other countries with small chemicals sectors but an interest in creating local chapters. For example, Peru's industry council has reportedly explored the possibility of establishing a similar agreement, but so far has not been able to find in Peru a willing and capable partner NGO.[8] The Ecuadorian case highlights the fact that this local organizational capacity can be gained and sustained through any number of institutions or arrangements.

RESPONSIBLE CARE IN CHILE

Conducta Responsable, Chile's Responsible Care chapter, was established in 1994, slightly later than its counterparts in Argentina and Brazil. As in those countries, transnational firms played important roles early on by introducing and promoting the regime. The support of the Chilean chemical manufacturers' industry council, ASIQUIM, for *Conducta Responsable* is reflected in its 1997 decision to make participation mandatory for all new members (though as of 2007, participation was still not required of all members).

The Chilean industry is relatively small and organized, and the industry association ASIQUIM plays an important role for the community. Chile's path of industrialization and liberalization during the 1980s and 1990s followed a corporatist model of state collaboration with organized business interests. Many of the negotiations over regulation, subsidies, liberalization, and so forth that took place over the last 20 years were between divisions of the government and these industry associations, so that remaining outside of a firm's relevant association was disadvantageous. As a result, business communities in Chile tend to have highly developed and active associations that play key roles in coordinating state-business relations.

This helps to explain why *Conducta Responsable* has a surprisingly high share—43 percent—of small-sized companies as members, well above the rates in Argentina or Brazil. Despite these high levels of formal participation, though, rates of active participation in the form of regularly submitted self-evaluations are roughly similar, at around 70 percent. Around 20 percent of Chilean member firms have undergone external auditing to verify their performances. These audits are conducted under a process similar to that used by the Canadian chemicals industry, which involves independent certifiers and community and NGO representatives, in addition to industry representatives.[9]

Conducta Responsable is exceptional because of the important partnership that the chapter enjoys with government agencies. Early in its development, in 1997, the chapter's administrators at ASIQUIM signed an agreement with the quasi-governmental Chilean Safety Association (*Asociación Chilena de Seguridad*), a national agency that advocates workers' safety. This partnership involves little more than the mutual sharing of information, technical assistance, and formal mutual endorsements. However, ASIQUIM emphasizes this cooperation, as well as other indications of support from the government and public agencies, to demonstrate the legitimacy of *Conducta Responsable*.

Chilean government agencies endorse *Conducta Responsable* and its principles in other ways. State cooperation includes the joint coordination of the chapter's annual national conference and celebration of the national "Conducta Responsable Day." These events culminate with the pronouncement of the winners of the annual Responsible Care prizes. Similar to competitions in Ecuador and Brazil, this contest among companies judged to have performed best in line with the objectives and codes of Responsible Care was initiated at the establishment of the chapter. As in Brazil, this state-endorsed competition is highly publicized and held in high regard by the Chilean business community.

Administrators of *Conducta Responsable* claim that as participation has increased, the regime has much improved the industry's relations with regulators. Environmental regulators have agreed to take firms' self-reporting and verification status into consideration when handling licensing and inspections and have accepted recommendations from chapter administrators and members in regard to technical standards. Moreover, participation in *Conducta Responsable* is also reported to be useful for firms seeking financial assistance from state-managed funds and development banks.[10]

Chilean industry officials and *Conducta Responsable* administrators highlight the chapter as one element of the national effort to comply with international environmental conventions. Through the program, and by cooperating actively in the international initiative Clean Production (*Producción Limpia*), the chemical industry in Chile claims to be at the forefront of the movement in that country for corporate environmental responsibility.

This case bolsters the support for the thesis that local organizational capacity is of paramount importance to the effectiveness of these global regimes. In Chile, the partnership between the industry council and various governmental and public agencies has been important to the chapter's success, particularly in a nation where state-business collaboration is the norm. Along with the FSC in Bolivia, this case calls into question the finding from the Argentine and Brazilian cases that government actors play minor roles in these regimes' development and effectiveness at the national level. Instead, in Chile the government has been a critical partner of the industry's *Conducta Responsable* chapter and has influenced its effectiveness in several ways.

In other respects this case supports our previous findings. Demand-side factors offer little explanation for why this chapter is more effective than that of Ecuador or Argentina, and less so than Brazil's Atuação Responsável. The support from transnational firms is similar in all cases and cannot easily be linked to the divergence in their outcomes. Instead, to explain this variance we must examine the local actors and interests, both within the industry and around the industry, that have an interest and capacity for supporting or administering a local regime chapter.

This comparative overview of global private regime chapters across South America highlights the diversity in the type of local institutional resources that can effectively support a local chapter. In the case of Brazil's Responsible Care chapter, local organizational capacity comes entirely from within the industry, via the industry association. In Ecuador's Responsible Care, it comes via a strategic partnership initiated by a local environmental NGO. In the case of Chile's Responsible Care chapter, local capacity consists partly of partnership with various government agencies and labor organizations. In Argentina's two chapters, this type of capacity has yet to develop. Argentina's Responsible Care and FSC chapters are stunted, due to indifferent industry councils and fragmented advocacy communities. The key point is that, across all these cases, demand-side factors in terms of foreign market demand and transnational company presence are roughly similar; what has determined regime success in each case is whether, and how, local organizations have been able to coordinate to make these regimes effective.

IMPLICATIONS FOR PRIVATE ENVIRONMENTAL REGIMES AS TOOLS FOR GLOBAL GOVERNANCE

To summarize, this study asks why global private environmental regimes, which purport to be independent of government action and are believed to depend in developing countries on demand from transnational firms and global supply chains, should vary so widely in their implementation at the national level. It seeks to explain observed differences in the effectiveness of two prominent private regimes, in two industries, as they have been implemented in two similar developing countries, Argentina and Brazil. Analysis of these four cases reveals patterns that contrast with the common wisdom regarding the factors that support and promote these private regimes at the level of national industries. There are four key findings.

First, demand-side factors such as market benefits, production chain pressures, or threats of tightening regulation have had a relatively small impact in these cases. Instead, variation in chapter effectiveness seems largely determined by whether local factors or conditions support or do not support the supply, at the local level, of these regimes. Private regimes are viable and effective in developing countries only to the extent that local actors (either individually or collectively) are able and willing to provide administration that serves the needs of members and responds to challenges and changing circumstances.

The second key finding is that although transnational firms and NGOs play significant roles in introducing these regimes to developing country associations and groups, as national chapters become operational the relevance of these foreign actors is eclipsed rapidly by that of local firms, NGOs, and networks. Over time, the impact of the efforts of transnational firms and NGOs to promote these regimes decreases, and the determinative factor is whether local firms and groups have been willing to advocate for and lead the chapter.

Third, government actors have less influence in developing states than in developed states over the growth and effectiveness of these private environmental regimes. Policies and actions aimed at promoting or opposing these national chapters have had little effect. In the Argentine cases, they have even been counterproductive.

This does not mean, however, that government policies are irrelevant to the success of these regimes. Cases in Bolivia and Chile support a fourth finding: that governments can indeed play key, direct roles as advocates or partners. Moreover, in Argentina and Brazil state actors are relevant indirectly, not as a direct source of incentive but as a key element of the institutional and environmental conditions that industries, NGOs, and other stakeholders face when managing or participating in these private regimes. In these cases the legacy of past state policies and patterns in the relations between state and industry actors help determine the local organizational capacities for the administration of regime chapters. The importance of institutional and political legacies for determining local organizational capacity is the fourth key finding.

In recent years, private actors in Western Europe and North America including firms, NGOs, certification agencies, and environmental advocates have created dozens of private regimes to promote more environmentally responsible industrial practice. Several of these programs, including Responsible Care and the Forest Stewardship Council, have spread globally. New private regulatory regimes emerge every year, aimed at encouraging various types of environmental, labor, and community relations practice.

The findings of this study suggest that despite their attempts to promote uniformity in norms and practices, these global regimes will vary in their effectiveness across developing nations. Although formally independent of the actions or inactions of national governments, these regimes depend on factors and conditions that pertain to the local societies, institutions, and economies in which they function, and over which they have no control. Furthermore, these factors and conditions are the legacies of decades of previous industrial, economic, and regulatory policies and cannot easily be changed even if government and private actors agree to that objective. As a result, local implementation of these global regimes is diverse and is likely to remain diverse. Within many—perhaps most—developing states, effective regime implementation may be impossible.

This suggests that the viability of these private regulatory regimes is limited, instead of universal. As instruments for the provision of global public goods in the form of effective and more efficient environmental regulation, these regimes are seriously flawed. If they can be expected to be effective in only a handful of industrialized democracies, mostly in Western Europe or North America (where environmental practice is already significantly regulated), they are hardly the tools that the world needs to curb the environmental harms of widespread industrialization.

Contrary to the common view, especially among their advocates, these global private environmental regimes are not neutral in their market effects. Firms, particularly nontransnational firms, do not enjoy equal access to these regimes. The industries they regulate are not equally predisposed to be amenable to their operations. Instead, firms operate within national industrial environments that are either advantageous or disadvantageous to regime success, and their quality is not amenable to alteration. For these reasons, global private regulatory regimes are not elements of a level playing field. Despite being designed to operate this way, cross-national and cross-industry differences at the local level create inconsistencies in their viability and effectiveness.

As this divergence in the effectiveness of national chapters increases, these global private regimes face an array of problems. Not just major firms, but labor unions and organized civil society groups (for example, associations of indigenous peoples in Canada) have learned to pursue their domestic and international interests through participation in these private regimes. Many firms and producers in Southern, developing countries, and their governments, remain concerned that these regimes serve the interests of their competitors in the North. As a result, producers in disadvantaged states or industries are likely to press on with their complaints within formal trade institutions: see, for example, the debate within the World Trade Organization over FSC as a nontariff import barrier.

As these dynamics unfold, private environmental regimes may lose their credibility in the eyes of retailers, consumers, environmental activists, and other stakeholders if they are perceived as strategic tools in the hands of larger firms and better-organized industry groups, rather than as win-win solutions enhancing the public welfare. Seeing that the effects of these private regimes are limited to specific countries, or segments of the industry, their advocates could withdraw their support and instead promote more traditional, state-based, universal forms of regulation.

How can administrators of global regimes address these challenges? This study indicates that the best means of doing so would be to focus on creating and expanding local organizational capacity in the form of additional institutional resources, partnerships, and coalitions upon which local administrators can draw. This contrasts with the current orientation of these global managers, who are largely focused on expanding formal membership (with less regard for the quality of participation) or building market share for certified products or member companies. Instead, advocates and administrators should be sharing among their national-level administrators successful cases of building local capacity, especially exceptional cases such as the Responsible Care chapter in Ecuador. They should also sometimes apply pressure within their own ranks for a more serious commitment to building and enhancing the regime at the local level.

This study demonstrates that to understand the impacts and potential of these private environmental regimes as instruments for global governance, we must look much more closely at their adoption and implementation by local parties. Unfortunately, its analysis suggests that these regimes' effectiveness may be more limited than many other studies would lead us to believe, because the conditions that constrain their growth are not easily overcome even by the most proactive and innovative advocacy tactics. This is not, however, meant to refute their importance. If these private regimes can thrive only in countries where conditions are favorable—such as in Brazil, India, and South Africa—that achievement alone is worth the dedication of their members and administrators. Furthermore, the lessons from these cases suggest there is ample ground for creative problem solving and coalition building, even under difficult circumstances. After all, the market and nonmarket environments in which these regimes formed and continue to grow are themselves constantly evolving in response to new pressures, technologies, governments, and crises. The better we understand the constraints they pose, and the sources of those constraints, the more effectively we can overcome them.

APPENDIX:
LIST OF INTERVIEWS

SEPTEMBER–NOVEMBER 2004

Regarding the forestry industry and the Forest Stewardship Council

Date	Type of organization, size, location, other relevant information
Nov. 11-12	Medium-sized Argentine company; Corrientes province; certified FSC; produces cut wood products
Nov. 9	Medium-sized Argentine company; Buenos Aires province; certified FSC; produces particle board and fiberboard
Nov. 15	Small consulting company; Misiones province; individual formerly managed ISO 14001 certification at large Argentine forestry company
Nov. 17	Small consulting and forestry services company; Entre Rios province; coordinates and manages FSC certification for local group of small producers; individual also manages environmental system for large forestry TNC (not certified)
Nov. 11	Large company; Misiones province; member of major Chilean industry group; not certified
Nov. 18	Medium Argentine company; Buenos Aires province; certified FSC; produces fiberboard
Nov. 11	Medium Argentine company; Misiones province; certified FSC; produces plywood and wood planks; major global producer of eucalyptus products
Nov. 18	Medium Argentine company; Buenos Aires province; certified FSC; produces fiberboard
Nov. 12	Large company; Corrientes province; member of major Chilean industry group; not certified
Nov. 19	Argentine chapter of a major environmental NGO
Nov. 4	National institute of standardization, coordinating initiative to create national forestry standard
Nov. 19	Private consultant to forest companies; worked previously for French forestry company; participates in working groups for national standards programs (FSC and IRAM)

Nov. 12	Professor of agricultural engineering in Posadas, Misiones province; participated in regional FSC initiative; previously coordinated FSC certification at local Argentine company
Sept. 29; Oct. 18	Environmental NGO that coordinates national FSC initiative; Buenos Aires
Nov. 22	Environmental NGO that coordinates national FSC initiative; Buenos Aires
Oct. 29	Federal agency that is organizing effort at national forestry standard
Nov. 17	Federal agency with programs that promote forestry certification
Nov. 1	National forest industry association
Oct. 21	Federal agency with programs that promote forestry industry modernization and native forest preservation

Regarding the chemicals manufacturing industry and Responsible Care

Oct. 13	Medium Dutch-owned company; Buenos Aires province; produces phosphorous and derivatives
Oct. 15	Major U.S. transnational corporation
Oct. 14	Small Argentine company; Buenos Aires province
Oct. 27	Medium, specialized unit of French transnational; Buenos Aires province
Oct. 13	Small, local unit of U.S. transnational
Oct. 12	Medium-sized company; subsidiary of major U.S. transnational corporation; Bahía Blanca petrochemical manufacturing center; Buenos Aires province
Sept. 27, Nov. 2	Argentine Responsible Care chapter; within national chemical industry association
Oct. 22	Argentine transport company; contractor to several chemical companies
Oct. 6	Federal agency administering "*Producción Limpia y Consumo Responsable*" program
Oct. 25	Attorney specializing in environmental law; several chemical company clients
Oct. 28	Provincial congressional advisor (Buenos Aires province) with expertise in environmental law and regulations, especially regarding chemical industry
Oct. 22	Major German transnational corporation

Regarding environmental certification systems in general

Nov. 2	Auditor, English certification services company
Nov. 23	National accreditation institute
Oct. 29	Auditor, English certification services company
Oct. 7	Former auditor; editor of journals on certification and on corporate responsibility
Nov. 2	Auditor, Argentine certification company
Nov. 5	Independent auditor and consultant; former executive at major transnational certification services company; former board member at national accreditation institute
Nov. 8	Auditor with national accreditation institute
Nov. 5	Auditor and training coordinator with Norwegian certification company

General interviews

Nov. 1	Federal agency for the promotion of industrial technology; environmental division
Nov. 4	International business association promoting sustainable development and corporate responsibility
Oct. 8	Research center on economic and social development
Oct. 20	U.S. Chamber of Commerce; department of promotion of corporate responsibility
Oct. 21	Argentine social and environmental development NGO
Oct. 7	Argentine NGO focused on social development and building corporate-NGO links
Sept. 29	Argentine environmental NGO; public-private projects division
Oct. 28	Professor of ecology and environmental law; advisor to Buenos Aires provincial government
Nov. 24	Attorney and journalist specializing in environmental law

BRAZIL

JUNE–AUGUST 2005

Regarding the forestry industry and the Forest Stewardship Council

Date	Type of organization, size, location, other relevant information
July 17	Staff person at FSC-Brazil, Brasilia
July 17	Director of FSC-Brazil, Brasilia
July 18	Official at major international environmental NGO, Brasilia
July 18	Official at major international environmental NGO, Brasilia
July 18	Official at the Forestry Division of the national environmental ministry, Brasilia
July 19	Official and expert on wood products market, consultant to a major international environmental NGO, Brasilia
July 20	Official at the office of the World Bank, Brasilia
July 21	Official at the national environmental ministry, Brasilia
July 25	Official at a Brazilian forestry NGO, Brasilia
Aug. 2	Large Brazilian forestry plantations and pulp and paper firm, operations nationwide, São Paulo
Aug. 3	National forestry industry society, coordinator of the national forestry standard program, São Paulo
Aug. 5	Independent auditor of forestry standards, São Paulo
Aug. 10	Director of Brazilian forestry certification services company, Piraçicaba, São Paulo
Aug. 10	Auditor of forests certification, Piraçicaba, São Paulo
Aug. 10	Independent auditor of forest certification, associated with major U.S. certification services firm, Piraçicaba, São Paulo

Aug. 12	Large Brazilian forest plantation and pulp and paper firm, operations throughout southeastern region and in Bahia, São Paulo
Aug. 16	Small Brazilian forest harvester, Amazonas state (phone interview)
Aug. 17	Small Brazilian forest plantation manager and native forest harvester, Rondônia state (Amazonia) (phone interview)
Sept. 6	Medium Brazilian forest harvester, native forest, in several Amazonia states (phone interview)

Regarding the chemicals manufacturing industry and Responsible Care

June 27	Staff person at Responsible Care, São Paulo
June 27	Director of Responsible Care, São Paulo
July 5	Medium Brazilian company, operations in São Paulo and Rio Grande do Sul, several product lines
July 7	Large Brazilian company, formerly state-owned, nationwide operations, multiple product lines
July 8	Large transnational company (U.S.), operations nationwide, multiple product lines
July 8	Officials at São Paulo state environmental regulatory agency, pollution control division
July 12	Staff person/auditor at Responsible Care Brazil, São Paulo
July 13	Auditor at São Paulo state environmental regulatory agency, pollution control division
July 14	Large transnational company (German), operations around the country, multiple product lines
July 15	Large Brazilian company; operations in São Paulo state and at Camaçari in Bahia, multiple product lines
July 26	Large transnational company (German), director of South American Responsible Care implementation
July 27	Independent auditor and consultant to Responsible Care Brazil, São Paulo
July 28	Large Brazilian company, formerly state owned, nationwide operations, multiple product lines
July 28	Medium Brazilian company, operations in Rio Grande do Sul, Paraná, and São Paulo, several product lines
July 29	Director of Responsible Care Ecuador
July 29	Director of Responsible Care Chile
Aug. 2	Large Brazilian/European jointly managed company, operations nationwide, several product lines
Aug. 3	Small Brazilian company, São Paulo state, manufactures paints and wood finishes
Aug. 11	Medium Brazilian company, operations in São Paulo and Rio de Janeiro, several product lines
Aug. 11	Large transnational company (U.S.), operations in São Paulo and at Camaçari, multiple product lines
Aug. 16	Director of Responsible Care Brazil, São Paulo
Aug. 16	Staff person/auditor at Responsible Care Brazil, São Paulo

General interviews

June 27	Attorney and assistant attorney general for the state of São Paulo, with extensive experience in environmental regulation
June 29	Journalist and editor at a national newspaper who covers the Brazilian business community, São Paulo
June 30	Professor at the University of São Paulo, department of industrial chemistry
July 4	Professor at the University of São Paulo, department of ecology and life sciences
July 5	Editor of regional journal on corporate environmental practice
July 29	Attorney involved in environmental issues, São Paulo
Aug. 16	Journalist at a national newspaper who covers environmental issues, São Paulo

NOTES

INTRODUCTION

1. The DuPont Corporation, for example, was instrumental in introducing Responsible Care in both countries, and the World Wildlife Fund from the beginning provided critical, ongoing support for the FSC.

CHAPTER 1

1. See Steven Krasner's edited volume *International Regimes* (1983).
2. For an excellent discussion of program categorization according to their degree of institutionalization, see the introduction in Cutler, Haufler, and Porter's edited volume (1999).
3. ISO's 14000 series of environmental management standards do not fit these criteria in two ways. ISO is a quasi-governmental organization with a membership made up of national standards agencies, and ISO's management standards are not specific to any particular industry, though they serve as the model for several industry-dedicated private regimes.
4. This categorical framework is based on Vinod Aggarwal's analysis of international trade regimes in *Liberal Protectionism: The International Politics of Organized Textiles Trade* (1985).
5. For a detailed discussion of nonstate, market-driven regimes see the introductory chapter in Cashore, Auld, and Newsome (2004a).
6. See the Forests and the European Union Resource Network's (FERN) publication *Echoes in the Forest* for a detailed discussion of standards types in regards to sustainable forest management. Available at: *http://www.fern.org*.
7. David Vogel's excellent volume *The Market for Virtue* (2005) provides a critical accounting of the CSR movement.
8. Ronie Garcia-Johnson and her colleagues at Duke University first presented this type of categorization at a University Symposium in 2001.
9. The seminal statement on the nature and importance of market-based social institutions belongs to Douglas North: *Institutions, Institutional Change, and Economic Performance* (1990).

10. For a thorough review of the ontological bases of the major schools of thinking on environmental politics, see Ronnie Lipschutz's volume *Global Environmental Politics: Power, Perspectives, and Practice* (2004).

11. See Tony Porter's contribution in the edited volume by Cutler, Haufler, and Porter (1999).

12. For example, see Jennifer Clapp's analysis of the lopsided negotiations of the ISO 14000 family of environmental management standards in the journal *Global Governance* (1998).

13. See, for example, the studies of NGO-business alliances in David Murphy and John Bendell's *In the Company of Partners* (1997), or the theoretically rich analysis of transnational advocacy networks in Margaret Keck and Kathryn Sikkink's book *Activists Beyond Borders* (1999).

14. John Braithwaite and Peter Drahos, for example, provide a comprehensive, provocative, and thoughtful road map for the establishment and strategic positioning of such local-global networks in their magisterial volume *Global Business Regulation* (2000).

15. Again, for a more detailed treatment see Murphy and Bendell (1997).

16. Ben Cashore, Graeme Auld, and Deanna Newsome compare the development of the Forest Stewardship Council in these three countries in *Governing Through Markets* (2004a).

17. Kenneth AbbottKen Abbot and Duncan Snidal (2006) describe a complete rule-making and enforcement cycle based upon the complementary competencies of key public and private actors.

18. Part of the allure of the Forest Stewardship Council, particularly in nations where the public has relatively little confidence in the effectiveness or fairness of their governments, is its open, democratic, consensus-based process for the establishment of national and local forestry standards. In nations such as Bolivia (see Chapter 7), where compliance with FSC standards is practically a legal obligation, the FSC's standards-writing councils indeed serve as quasi-governmental bodies.

19. See Kathryn *Rules for the Global Economy* and Harrison's "Talking with the Donkey" (1999) for a detailed appraisal of these relationships.

20. See Kate O'Neill's (2004) overview of the literature on global private environmental governance.

21. See Ben Cashore's article "Legitimacy and the Privatization of Environmental Governance" in the journal *Governance* (October 2002).

22. Forest Reinhardt's book *Down to Earth* (1999) surveys the market and non-market calculations managers must make when deciding if and to what degree their firm should participate in green regimes, and presents several case studies of firms engaged in various types of "greening."

23. Aseem Prakash probes these management decisions in depth in his book *Greening the Firm: The Politics of Corporate Environmentalism* (2000a).

24. For a multinational examination of the relative importance of these factors, using mostly ISO 14001 certifications, see Prakash and Potoski (2006).

CHAPTER 2

1. For examples of these inquiries into the effects of international institutions, see Stein (1982), Aggarwal (1998), and Aggarwal and Dupont (1999).

2. For a thoughtful discussion of this shortcoming of modern institutional approaches to international relations, see Simmons and Martin (1998).

3. These challenges, discussed in detail by Haas, Keohane, and Levy (1993), continue to frustrate research on global environmental politics.

4. This formulation is based on Thomas Bernauer (1995), with the additional criterion number 4 from a constructivist perspective.

5. This is an approximation of the prescription that Haas, Keohane, and Levy (1993) offer on page 7 of their early work on international environmental institutions. Their focus, however, was on the behaviors of governments as signatories to an international regime, not on producers.

6. This approach is especially common in studies of the effectiveness of international regimes, perhaps because it is relatively simple to apply in evaluating compliance across a homogenous membership: states.

7. The four cases are the Forest Stewardship Council in Argentina, the Forest Stewardship Council in Brazil, Responsible Care in Argentina, and Responsible Care in Brazil.

8. For example, we can compare the costs and difficulty of FSC certification of large plantation operators against those of small private landowners by considering the data in each case in terms of the budgets and other resources of each. Comparing data across different types or sizes of producers is a significant challenge to other methods of measuring effectiveness, particularly those based on quantitative estimates.

9. High levels of participation also indicate a less conspicuous aspect of effectiveness. In private regulation, participation is voluntary. Broad membership in a private environmental regime reflects, and at the same time reinforces, its legitimacy. When a majority of an industry commits itself voluntarily to a regime, this suggests the acceptance of the regime's principles and its benefits for members and the broader public.

   Legitimacy in this sense refers to *internal* legitimacy bestowed upon a regime or system by its members and does not include the *external* legitimacy that must come from actors and observers on the outside.

10. As discussed earlier in this chapter, actual impact on the environment is difficult to measure or interpret because of the complexity of environmental factors at play. In contrast, it is a relatively simple matter to observe modifications in participants' practices, which are often formalized in standard procedures and policies, management and training guidelines, and/or the use of new technologies that improve efficiency.

11. Aseem Prakash (2000b) takes a similar approach in his analysis of the U.S. chapter of Responsible Care.

12. See David Vogel's (1995) book *Trading Up*.

13. See Gereffi and Korzeniewicz (1994), and for a more recent application of the same framework see Jeppesen and Hansen (2004).

14. For a comprehensive discussion of the environmental impact of consumer-based pressures, and the strategies by which different companies have responded, see Reinhardt (2000).

15. Several studies have explored the impact of national regulatory systems, or cultures, on the strategies of firms, including their propensity to form private regimes. Among the best of these is Gunningham, et al. (2003), which examines the impact of environmental laws on pulp-and-paper production practices across different regulatory jurisdictions. Prakash and Potoski (2006) make a similar argument based upon national-level data across several OECD member nations.

16. Cashore, Auld, and Newsom's book *Governing Through Markets* (2004a) traces these processes in great detail in the case of the Forest Stewardship Council's development in the United States, Canada, Germany, the UK, and Sweden.

17. In fact, our comparison suggests that a back-of-the-envelope correlation exists, positively, between the number of staff in a national regime chapter's administrative organization and that regime's effectiveness at managing these challenges.

18. See again Cashore, Auld, and Newsom (2004a).

19. See, for example, Ronie Garcia-Johnson's (2000) emphasis on the advocacy roles played by U.S. industry leaders in her study of the Responsible Care initiative in Mexico and Brazil, and Gereffi, Garcia-Johnson, and Sasser (2001), who focus on transnational advocacy groups and firms as the key leaders of the movement toward global corporate social responsibility.

20. See, in particular, Aseem Prakash's emphasis on industry concentration as a major explanation for the success or failure of Responsible Care and other regimes, modeled as club goods, or Prakash and Potowski's (2005) focus on levels of industry concentration as a major determinant of effective collective action.

21. However, under the Kirschner administrations, Argentina has tangled with foreign creditors while still hoping to attract foreign investment.

CHAPTER 3

1. United Nations Food and Agriculture Organization. (2000). *Global Forest Resources Assessment 2000*. FAO Forestry Papers 140. Available at: *http://www. fao.org/documents*.

2. The nations of Brazil, Indonesia, Sudan, Myanmar, Zambia, United Republic of Tanzania, Nigeria, Democratic Republic of the Congo, Zimbabwe, and Venezuela (Bolivarian Republic of) had a combined net forest loss of 8.2 million hectares per year in 2000–2005. See *http://www.fao.org/forestry*.

3. United Nations Food and Agriculture Organization (FAO). (2006). *Global Forest Resources Assessment 2005*. Available at: *http://www.fao.org/forestry/site/ fra2005/en/*

4. United Nations FAO. (2006).

5. Dauvergne, Peter. (2005). The Environmental Challenge to Loggers in the Asia-Pacific: Corporate Practices in Informal Regimes of Governance.

In David L. Levy and Peter J. Newell, eds. *The Business of Global Environmental Governance.* Cambridge: The MIT Press.

6. *Folha de São Paulo.* August 18, 2005. "Operacão Curupira prende 16 pessoas por extração illegal de Madeira." Available at: *http://www1.folha.uol.com.br*

7. *http://www.unctad.org/en/docs/tdtimber3d12_en.pdf*

8. FSC. *FSC Certification: Maps, Graphs, and Statistics,* April 2008. Available at *www.fsc.org.*

9. FSC's system is currently based on ten global principles of sustainable forestry, which are implemented through the application of 55 performance measures. See the FSC's *Principles and Criteria for Forest Stewardship,* at *www.fsc.org.*

10. Information on the PEFC is available at: *http://www.pefc.org/internet/html/index.htm.*

11. House of Commons Environmental Audit Committee. (2005). Sustainable Timber: Second Report of Session 2004–2005. Available at: *www.parliament.uk/parliamentary_committees/environmental_audit_committee.cfm*

CHAPTER 4

1. Setting aside, for the purposes of this study, non-Western or anticapitalist movements and networks, or criminal groups, which have also spread with globalization.

2. In contrast, in Argentina (and in other regions of Brazil), small-scale, privately owned forests still account for a significant share of regional wood supplies. See Sánchez-Acosta (2000).

3. Interviews with officials at certification agencies, August 11, 2005; at the World Bank office in Brasilia, July 20, 2005; at a national industry association, August 3, 2005; at WWF-Brazil and at FSC-Brazil, July 25, 2005.

4. For example, Northern groups promoted FSC's decision to replace, in its fiberboard and composite board certification, the logo "70 percent FSC Certified" with "FSC Mixed Sources" to help European producers qualify for the logo. Brazil's large, extremely efficient producers can provide internally 100 percent or 90 percent FSC-certified fiberboard more easily than can European producers, who generally must purchase fiber from small producers.

5. For a good overview of the environmental regulatory structures in the region, see Kathryn Hochstetler's (2003) essay "Fading Green? Environmental Politics in the Mercosur Free Trade Agreement."

6. For an in-depth analysis of this unusual regulatory instrument, see Lesley McAllister's (2004) dissertation *Environmental Enforcement and the Rule of Law in Brazil.*

7. Interviews with officials at the environmental division of the Health Secretariat, October 6, 2004, and an official at the national institute for standardization, November 17, 2004.

8. Existing certifications meet generic FSC international standards.

9. Sánchez Acosta (2000).

10. Braier (2004); República Argentina (2002).

11. Bolivia (17 forests, 1,727,104 hectares), where FSC certification is required of all public forest concessions, is an interesting exception. The Bolivian FSC case is discussed in detail in Chapter 7.

12. According to the 2006 Annual Yearbook of BRACELPA, the Brazilian Cellulose and Paper Association.

13. Interview with an official at a major tree plantation company, August 16, 2005.

14. See Gunningham, et al. (2003).

15. Interview with an expert on wood markets at the WWF, July 21, 2005, and certification agency official on August 11, 2005.

16. Of course this is due partly to the vastness of Brazil's native forests. Interview with a certifier affiliated with an international forest certification agency, August 11, 2005.

17. Interview with director of the forestry division at the Ministry of the Environment, July 19, 2005.

18. Interviews with an official at a large Chilean-owned company in Corrientes province, and an official at a federal agency that promotes forestry certification, November 12 and 17, 2004.

19. In order to comply, one company was forced to build a special storage unit for the indefinite storage of packaging from chemicals used, because neither waste disposal services nor local chemical companies offered the type of treatment process required by law (interview with an official at a forest services company in Corrientes province, November 17, 2004). In the province of Corrientes, compliance with local law included registration on a government list that did not exist. Two years after the provincial government created a registry to accommodate the needs of these companies, as demanded by FSC, those two companies were still the only ones in the registry (interview with an official at an FSC-certified company, November 11, 2004).

20. Interviews with an official at a major Chilean-owned company in Misiones province, and a professor of agricultural engineering in Misiones province, November 11 and 12, 2004.

21. All FSC certifications in Argentina have been and are audited by one of three agencies: SGS Certification Services, Inc., Scientific Certification Systems, and the nonprofit *Smartwood* program of the Rainforest Alliance.

22. Interview with a professor of agricultural engineering in Misiones province, November 12, 2004.

23. The most significant change in environmental practices that FSC certification demands of Brazilian plantation operators is minimization of the use of chemicals. Looking ahead, industry officials predict that the next great hurdle will be FSC's planned prohibition of transgenic technology. Since most tropical tree farms grow genetically modified strains of eucalyptus and pine, this new rule may prove extremely costly and severely reduce their international competitiveness. For this reason, some managers and industry officials in Brazil view FSC as potentially a competitive tool that their Northern competition may use for market protection, an issue to which we will return later in this chapter.

24. Interviews with officials of two certification agencies and managers of two firms that operate in native forest, August 11 and 23, and September 17, 2005.

25. Professor of environmental studies interviewed in Pousadas, Misiones Province, November 12, 2004.

26. Interviews with an official at an FSC-certified company and an official at a federal agency that promotes forestry certification, November 11 and 17, 2004.

27. Interview with an official at an FSC-certified company.

28. One official explained that the sale of specialized high-end products to the Home Depot was the firm's chief rationale for seeking FSC certification in the first place. Interviews with an official at an FSC-certified company and with the coordinator of the FSC, November 18 and September 29, 2004.

29. Interview with an industry consultant in Misiones province, November 15, 2004.

30. Interview with an expert on forest goods markets with the WWF, July 24, 2005.

31. Interview with an official at a major tree plantation firm, August 16, 2005.

32. Interview with World Bank official, July 25, 2005, and with a manager of a major pulp and paper firm, August 16, 2005.

33. Phone interview with an official at a Brazilian forestry firm operating in the Amazon, September 16, 2005.

34. Interview with an official at a major Brazilian forest plantation firm, August 16, 2005.

35. Ibid.

36. Interview with FSC expert on wood and wood products markets, July 21, 2005.

37. Direct campaigns by Greenpeace and other NGOs tend to target exports and purchasers of illegal tropical wood in the United States or Europe.

38. Interview with an official from a major plantation firm, August 16, 2005.

39. Interview with an official at Greenpeace Argentina, November 19, 2005.

40. Director of a Brazilian environmental certifications services company, November 17, 2005.

41. Interviews with a certification agency official, August 11, 2005, and with an official at the World Bank, July 25, 2005.

42. Interviews with a certification agency official, August 11, 2005, the director of the forestry division at the Ministry of the Environment, July 19, 2005, and a World Bank official, July 25, 2005.

43. Interview with a certification agency official, August 11, 2005.

44. Interview with a World Bank official and board member of FSC-Brazil, July 25, 2005.

45. These include the WWF-Brazil, Greenpeace, Friends of the Earth, IMAFLORA (a Brazilian forestry services and certification firm), and Scientific Certification Systems.

46. Interview with officials at FSC-Brazil, July 18, 2005.

47. An investigative action by Greenpeace sparked the withdrawal of a producer's FSC certification, as well as that of the certifying agency that had certified the producer. Interview with an official at a certification agency, August 11, 2005.

48. Interview with WWF-Brazil staff in Brasilia, July 18, 2005, and with an official at the World Bank, July 25, 2005.

49. Interviews with officials at certifying agencies, August 11, 2005, a national industry association, August 12, 2005, and with managers at tree plantation firms, August 15 and 16, 2005.

50. Interviews with the official at the SAGPyA Forestry Division coordinating the national forestry standards initiative and with an official at the national institute for standardization, October 29 and November 4, 2004.

51. Interview with the director of FSC-Argentina, November 22, 2004.

52. Interview with an official at a federal research agency that promotes forestry certification, November 17, 2004.

53. Interview with an attorney and journalist who specializes in environmental law, November 24, 2004.

54. In June 2005, a federal sting operation netted 47 IBAMA officials in seven states profiting from illegal logging in the Amazon region. *Estado de São Paulo*, Ciência e Meio Ambiente, June 3, 2005: "PF prendeu pelo menos 95 na Operação Curupira."

55. Interview with an official at a certifying agency, August 11, 2005.

56. Interview with an official at the Ministry of the Environment, July 22, 2005, a small company operating in the Amazonian region, September 23, 2005, and two forest management certifiers, both on August 11, 2005.

57. Incidentally, quite unlike the case in the United States, where the SFI has little credibility outside of the industry. Interviews with officials at forestry companies August 16, 2005, a national industry association August 12, 2005, and certifying agencies, August 11, 2005.

58. Interview with an official at the Brazilian Forestry Society (SBS, Sociedade Brasileira de Silvacultura), August 12, 2005.

59. Interviews with certifying officials, August 11, 2005, officials at WWF-Brazil, July 18, 2005, and with firms with operations in Amazon native forests, August 17 and September 23, 2005.

60. In essence, a producer requires three types of licenses to operate. These come from governments (regulatory license), from other firms (economic license), and from the society (social license). The relevance of these varies across industries and issue areas, but firms that operate in environmentally sensitive industries, like forestry firms, require all three. See Neil Gunningham et al., *Shades of Green* (2003).

CHAPTER 5

1. Integrated production signifies the capacity of a company to control several, if not all, stages of production within itself, instead of contributing only one part of the production process. For example, a major chemicals manufacturer often has control, within the company or among its subsidiaries, over the extraction of crude oil from the ground, its delivery to a refinery, and its refinement into hundreds of different products, and even the further refinement of some of those goods into final products, such as cosmetics, pharmaceuticals, or plastics.

2. See Smart (1992), pages 70–71, for an overview of this failed campaign.

3. See Hoffman (1997).

4. Full information on the regime and its growth is available at the Web site: *http://www.responsiblecare.org*.

5. For an explanation of first-, second-, and third-party environmental regimes, see pages 16 and 17 in Chapter 1 of this volume, and Garcia-Johnson et al. (2000).
6. See Andrew King and Michael Lennox (2000).
7. This case is discussed in greater detail in Chapter 6.
8. Although, as Chapter 7 will discuss, Ecuador's exceptional RC chapter is partly administered by an environmental NGO, so that its self-reporting and verification system feature unique elements of control from outside the industry.

CHAPTER 6

1. See Pedro Wongtschowski's excellent 2002 survey of the Brazilian chemicals manufacturing sector, p. 150.
2. The Argentinian political economists Chudnovsky and Lópes provide a thorough examination of this record of deficient industrial policies in their 1997 book.
3. Again, the analysis found in Wongtschowski (2002) is highly recommended. On this point in particular, see p. 163–164.
4. For a more in-depth explanation of this process, see Schorr (2004) and Chudnovsky and Lópes (1997).
5. See Chudnovsky and Lópes (2001).
6. For a more in-depth discussion on Brazil's environmental regulatory system, read Kathryn Hochstetler's (2002) article.
7. For more information about how environmental policy and its implementation are conducted within Brazil's federalist system, read Seroa da Matta's excellent (2002) article.
8. Interview with the coordinator of CRM, September 27, 2004.
9. "Fue clausurada una petroquímica," *La Nacion*, April 24, 1992.
10. Interview with the coordinator of CRM, September 27, 2004.
11. Interviews with an official at a U.S.-owned chemical firm October 15, 2004, and with the coordinator of CRM, September 27, 2004.
12. Each round of self-evaluations included a minimum of six codes, or different areas, of evaluation. Each code was administered separately. At times a round took as long as three years to complete, so that different codes within the same round were often reported on by different sets of members, as some firms joined and others dropped from the program. For this reason, figures per round were averaged across these codes and the years covered are stated per round.
    Note: The 2007 international Responsible Care status report states that, in 2007, 67 percent of CIQyP's members were participating in CRM.
13. Interviews with the coordinator of CRM, an environmental manager at a Dutch company, and an environmental manager at a U.S. company, September 27, October 13, and October 12, 2004.
14. Interviews with an official at a major U.S. company and an environmental manager at a major German company, October 15 and October 22, 2004.
15. Interviews with an environmental manager at a French company, an official at an Argentine transport company, and an environmental manager at a major German company, October 22 and October 27, 2004.

16. Interview with an environmental manager at a major German company, October 22, 2004.

17. This information comes from ABIQUIM's 2004 annual report, available at https://www.abiquim.com.br.

18. Interviews with two ABIQUIM officials, June 27 and August 15, 2005.

19. This *VerificAR* system is discussed in detail further on in this chapter. Interviews with ABIQUIM officials and with officials from participating firms, June 29, July 4, July 13, and July 29, 2005.

20. Interview with officials at several chemical firms, July 13, August 3, and August 11, 2005.

21. Interview with ABIQUIM official, July 4, 2005.

22. Interview with ABIQUIM officials and a company manager, June 29, July 4, and August 3, 2005.

23. Interview with an environmental manager at a French company, October 27, 2004.

24. Interview with the coordinator of CRM, November 2, 2004.

25. Interview with an official at an Argentine company, October 14, 2004.

26. Interview with an official at a subsidiary of a U.S. company, October 13, 2004.

27. Interview with an environmental manager at a subsidiary of a Dutch company, October 13, 2004.

28. Interviews with an official at a U.S. company and the coordinator of CRM, October 12 and September 27, 2004.

29. Interview with the coordinator of CRM, September 27, 2004.

30. Interviews with company officials, August 1, August 3, and August 10, 2005.

31. The improvement came following the design and implementation of a closed system to capture and remove arsenic acid from the production process, instead of releasing it. This is according to an interview with the environmental manager at the company, on August 3, 2005.

32. The petrochemical center in Camaçari, Bahia state, for example, features the services of an environmental and safety services company. This firm oversees the practices of all the firms operating at the center, collects and disseminates best practices, coordinates community information and outreach programs, and facilitates relations with local regulators. Industry officials and regulators from across Brazil speak highly of the type of public-private collaborative operations in place at Camaçari (Calheira Barbosa 2003).

33. Interviews with company officials, July 13 and July 15, 2005.

34. Interview with a company official, July 13, 2005.

35. Interview with a company official, July 15, 2005.

36. Interview with an official at a major U.S. transnational, October 15, 2004.

37. Interview with an official at a major U.S. transnational, October 15, 2004.

38. Interviews with officials at a major U.S. transnational and a French firm, and with the coordinator of CRM, October 15 and 27, and September 27, 2004.

39. Interviews with officials at a French firm and a major U.S. transnational, October 27 and 15, 2004.

40. Interviews with officials at a major U.S. transnational and an Argentine firm, October 15 and 14, 2004.

41. Interviews with officials at an Argentine firm, a French firm, and a U.S. firm, October 14, 27, and 12, 2004.
42. Interview with ABIQUIM officials, July 4, 2005.
43. Interview with company official, July 13, 2005.
44. This reaction is atypical. Most large chemical firms with diverse product lines report having ISO 14001 certification, OSHA 18001 certification, and others in addition to participating in RC, in order to satisfy clients. In this case, this firm had committed to using RC as its global platform for safety, health, and environmental management, and wished to see RC strengthened instead of changing its internal management system.
45. Interview with ABIQUIM officials, July 4, 2005.
46. Interview with ABIQUIM officials, executive council members, and company managers, June 29, July 13, and July 29, 2005.
47. Interview with industry officials, July 13 and 15, 2005.
48. Interview with industry officials, June 22, August 8 and 17, 2005.
49. Interview with chemical industry officials, July 13, 23, and August 8, 2005.
50. For more detailed information on the environmental management practices, and Atuação Responsável, at the Camaçari complex, see Calheira Barbosa (2003).
51. Interview with ABIQUIM officials on July 4, 2005, and with officials at CETESB, the São Paulo state environmental protection agency, June 30, 2005.
52. Interview with a company official, August 11, 2005.
53. Interviews with officials from Dutch, Argentine, and U.S. firms, October 13, 14, and 15, 2004.
54. The committees are: Executive, Technical, Partners, Emergency Response, Quality, SASSMAQ (Transport partner certification), Human Resource Development, Community Dialogue, Product Management, Environment, Industry Protection, Worker Health and Safety, Process Security, Supplementary Goods, and Transportation.
55. For more detailed information about the progressive environmental management system in place at the Camaçari industrial center, see the report by Calheira Barbosa (2003).
56. Interviews with company officials on June 22, July 28, July 29, and August 3, 2005, and with an official at an NGO that monitors the chemical industry, July 5, 2005.
57. Interview with an AR official at ABIQUIM, July 29, 2005.
58. These programs are Producción Limpia ("Clean Production"), an initiative of the Secretary of the Environment begun in early 2004, and the Environmental Program of the National Institute of Industrial Technology (INTI).
59. Interviews with the coordinator of the CRM, and an official at a federal agency administering an antipollution program, September 27 and October 6, 2004.
60. Interview with the coordinator of Producción Limpia, October 6, 2004.
61. See Calheira (2003).
62. Interviews with company officials and ABIQUIM officials on June 29, July 13, July 27, July 28, July 29, and August 16, 2005.

63. For a complete description and analysis of the corporate social responsibility legacy within Brazil, see Cappellin and Giuliani (2002).
64. Garcia-Johnson (2000), for example, highlights the importance of U.S. transnational firms as channels for the diffusion of environmental regulation.

CHAPTER 7

1. Percentage obtained using data from the FAO's Global Forest Resources Assessment (2000), and the FSC International.
2. Fundación Vida Silvestre (2002).
3. For a detailed analysis, see Quevedo (2005), and the report by Fundación Vida Silvestre (2002).
4. Jack (1999) provides an illuminating analysis of the unusual circumstance of the FSC in Bolivia. See also an October 6, 2005, article, "What's New? Bolivia Takes the Lead in Smartwood/FSC Forest Certification" at the Rainforest Network Web site: *http://www.rainforest-alliance.org/news/2005/bolivia.html*.
5. Interview with the director of Responsabilidad Integral on July 26, 2005, in São Paulo, Brazil.
6. Interview with the director of Responsabilidad Integral on July 26, 2005, in São Paulo, Brazil.
7. See Responsible Care's (2007) annual report, and its section on the Ecuadorian chapter, for more details. This report is available at: *www.responsiblecare.org*.
8. Interview with the director of Responsabilidad Integral on July 26, 2005, in São Paulo, Brazil.
9. Interview with the director of Conducta Responsable on July 27, 2005, in São Paulo, Brazil.
10. Interview with the director of Conducta Responsable on July 27, 2005, in São Paulo, Brazil.

BIBLIOGRAPHY

ABIQUIM (Associação Brasileira da Indústria Química). 2005. *Relatório de Atuação Responsável 2005*. São Paulo: ABIQUIM.

ABIQUIM (Associação Brasileira da Indústria Química). 2004. *Relatório Annual 2004: A Abiquim e a Indústria Química Brasileira*. São Paulo: ABIQUIM.

Abbott, Kenneth W. and Duncan Snidal. 2006. "The Governance Triangle: Regulatory Standards Institutions and the Shadow of the State." Paper prepared for the Global Governance Project, Oxford University. October 2006.

Aggarwal, Vinod K. 1998. *Institutional Designs for a Complex World: Bargaining, Linkages, and Nesting*. Ithaca, NY: Cornell University Press.

———. 1985. *Liberal Protectionism: The International Politics of Organized Textiles Trade*. Berkeley: University of California Press.

Aggarwal, Vinod K. and Cedric Dupont. 1999. "Goods, Games, and Institutions". *International Political Science Review* 20 (4) October: 393–409.

Anderson, C. Leigh and Robert A. Kagan. 2000. "Adversarial Legalism and Transaction Costs: The Industrial-Flight Hypothesis Revisited." *International Review of Law and Economics* 20: 1–19.

Barkin, David. 2002. "The Greening of Business in Mexico", In Peter Utting (ed.), *The Greening of Business in Developing Countries: Rhetoric, Reality and Prospects*. London: Zed Books.

Baron, David P. 1996. *Business and Its Market and Nonmarket Environment*. Fourth Edition. New York: Prentice Hall.

Bartley, Tim. 2003. "Certifying Forests and Factories: States, Social Movements, and the Rise of Private Regulation in the Apparel and Forest Products Fields." *Politics and Society* 31 (3): 433–464.

Berman, Jonathan E. and Tobias Webb. 2003. *Race to the Top: Attracting and Enabling Global Sustainable Business*. Washington, DC: The World Bank.

Bernauer, Thomas. 1995. "The Effect of International Environmental Institutions: How We Might Learn More". *International Organization* 49 (2) Spring: 351–377.

Bernstein, Steven. 2000. "Globalization, Four Paths of Internationalization and Domestic Policy Change: The Case of Eco-Forestry Policy Change in British Columbia, Canada". *Canadian Journal of Political Science* 33 (1): 67–99.

Bernstein, Steven and Benjamin Cashore. 2005. "The Two-Level Logic of Non-State Market-Driven Global Governance." Unpublished paper shared with author.

Braier, F. Gustavo. 2004. *Informe Nacional: Argentina. Proyecto Información y Analises para el Manejo Forestal Sustenible.* Santiago: European Union/Food and Agricultural Organization.

Braithwaite, John and Peter Drahos. 2000. *Global Business Regulation.* Cambridge: Cambridge University Press.

Calheira Barbosa, Aurinézio. 2003. *Responsibilidade Social Corporativa de Pólo Industrial de Camaçari: Influência do Conselho Comúnitario Consultivo.* Dissertation for the Masters Program in Environmental Management at the Federal University of Bahia, Polytechnic School. Salvador, Bahía.

Cappellin, Paola and Gian Mario Giuliani. 2002. "The Political Economy of Corporate Social and Environmental Responsibility in Brazil." Paper written for the United Nations Research Institute for Social Development, Rio de Janeiro, 2002.

Cashore, Benjamin, F. Gale, E. Meidinger, and D. Newsom (eds). 2005. *Confronting Sustainability: Forest Certification in Developing and Transitioning Countries.* New Haven, CT: Yale School of Forestry and Environmental Studies Press.

Cashore, Benjamin, Graeme Auld and Deanna Newsom. 2004a. *Governing Through Markets: Forest Certification and the Emergence of Non-State Authority.* New Haven, CT: Yale University Press.

———. 2002. "Legitimacy and the Privatization of Environmental Governance: How Non-State Market-Driven (NSMD) Governance Systems Gain Rule-Making Authority". *Governance: An International Journal of Policy, Administration, and Institutions* 15 (4) October.

Chudnovsky, Daniel and Andrés Lópes. 2001. *La Transnacionalización de la Economía Argentina.* Buenos Aires: CENIT/EUDEBA.

——— (eds). 1997. *Auge y Ocaso del Capitalismo Asistido: la Indústria Petroquímica Latinoamericana.* Buenos Aires: Alianza Editorial.

Clapp, Jennifer. 1998. "The Privatization of Global Environmental Governance: ISO 14000 and the Developing World". *Global Governance* 4 (3): 295–316.

Corbett, Charles J. and Michael V. Russo. 2001. "The Impact of ISO 14001: Irrelevant or Invaluable?", In *ISO Management Systems,* Journal of the International Organization for Standardization, Geneva, Switzerland.

Cutler, A. Claire. 2002. "Private International Regimes and Interfirm Cooperation", In Rodney Bruce Hall and Thomas J. Biersteker (eds), *The Emergence of Private Authority in Global Governance.* Cambridge: Cambridge University Press.

———, Virginia Haufler and Tony Porter (eds). 1999. *Private Authority and International Affairs.* Albany: State University of New York Press.

Dauvergne, Peter. 2005. "The Environmental Challenge to Loggers in the Asia-Pacific: Corporate Practices in Informal Regimes of Governance", In David L. Levy and Peter J. Newell (eds), *The Business of Global Environmental Governance.* Cambridge, MA: MIT Press.

Delmas, Magali. 2002. "The Diffusion of Environmental Management Standards in Europe and the United States: An Institutional Perspective". *Policy Sciences* 35: 91–119.

——— and Michael W. Toffel. 2004. "Stakeholders and Environmental Management Practices: An Institutional Framework." *Business Strategy and the Environment* 13: 209–222.

Desai, Uday (ed.). 1998. *Ecological Policy and Politics in Developing Countries: Economic Growth, Democracy, and Environment.* Albany: State University of New York Press.

Erber, Fabio. 1997. "Desarrollo y Reestructuración de la Petroquímica Brasileira", In Daniel Chudnovsky and Andrés López (eds), *Auge y Ocaso del Capitalismo Asistido: La industria petroquímica latinoamericana.* Buenos Aires: Alianza Editorial.

Evans, Peter. 1997. "The Eclipse of the State? Reflections on Stateness in an Era of Globalization." *World Politics* 50 (October): 62–87.

——. 1995. *Embedded Autonomy: States and Industrial Transformation.* Princeton, NJ: Princeton University Press.

——. 1979. *Dependent Development: The Alliance of Multinational, State, and Local Capital in Brazil.* Princeton, NJ: Princeton University Press.

Falkner, Robert. 2003. "Private Environmental Governance and International Relations: Exploring the Links". *Global Environmental Politics* 3 (2) May.

FAO (UN Food and Agricultural Organization). 2002. *Socio-economic Trends and Outlook in Latin America: Implications for the Forestry Sector to 2020.* Prepared by Sandra Velarde. Available at: *http://www.fao.org/documents.*

——. 2000. *Global Forest Resources Assessment 2000.* FAO Forestry Papers 140. Available at: *http://www.fao.org/documents.*

Føllesdal, Andreas, Michele Micheletti, and Dietlind Stolle (eds). 2004b. *Politics, Products, and Markets: Exploring Political Consumerism Past and Present*, New Brunswick, NJ: Transaction Press at Rutgers University.

Forests and the European Union Resource Network (FERN). 2004. *Footprints in the Forest: Current Practices and Future Challenges in Forest Certification.. http://www. fern.org/media/documents/document_ 1890_1900.pdf.*

Finnemore, Martha and Kathryn Sikkink. 1998. "International Norms and Political Change." *International Organization* (Autumn): 887–917.

Fundación Vida Silvestre. 2002. *Informe: La Certificación y el Futuro de los Bosques.* Report on FSC available at: *http://www.vidasilvestre.org.ar/pdfs/FSC-informe.doc*

Garcia-Johnson, Ronie, Gary Gereffi and Erika Sasser. 2000. "Certification Institution Emergence: Explaining Variation." Unpublished paper from the Duke University *Symposium on Certification Institutions and Private Governance*, December 2001.

——. 2000. *Exporting Environmentalism: U.S. Multinational Chemical Corporations in Brazil and Mexico.* Cambridge, MA: MIT Press.

Garlipp, Rubens C. 2004. *Recursos Forestales-Brasil. Proyecto Información y Analises para el Manejo Forestal Sustenible.* Santiago: European Union/Food and Agricultural Organization.

George, Alexander L. and Andrew Bennett. 2005. *Case Studies and Theory Development in the Social Sciences.* Cambridge, MA: MIT Press.

—— and Timothy J. McKeown. 1985. "Case Studies and Theories of Organizational Decision Making". *Advances in Information Processing in Organizations* 2: 21–58.

Gereffi, Gary, Ronie Garcia-Johnson, and Erika Sasser 2001. "The NGO-Industrial Complex," *Foreign Policy* 125 (July–August): pp. 56–65.

Gereffi, Gary and Miguel Korzeniewicz. 1994. *Commodity Chains and Global Capitalism.* New York: Praeger Publishers.

Gereffi, Gary, and Donald L. Wyman (eds). 1990. *Manufacturing Miracles: Paths of Industrialization in Latin America and East Asia*. Princeton, NJ: Princeton University Press.

Goldstein, Judith and Roberto O. Keohane (eds). 1993. *Ideas and Foreign Policy: Beliefs, Institutions, and Political Change*. Ithaca, NY: Cornell University Press.

Gordon, Kathryn. 1999. "Rules for the Global Economy: Synergies between Voluntary and Binding Approaches." Organisation for Economic Co-operation and Development (OECD), Working Papers on International Investment, No. 1999/3.

Guimarães, Roberto P. 1991. *The Ecopolitics of Development in the Third World: Politics and Environment in Brazil*. Boulder, CO: Lynne Rienner Publishers.

Gunningham, Neil, Robert A. Kagan and Dorothy Thornton. 2003. *Shades of Green: Business, Regulation, and the Environment*. Stanford, CA: Stanford University Press.

Gunningham, Neil, Peter Grabowsky, and Darren Sinclair. 1999. *Smart Regulation: Designing Environmental Policy*. Oxford: Oxford University Press.

Haas, Peter. 1989. "Do Regimes Matter? Epistemic Communities and Mediterranean Pollution Control." *International Organization* 43: 377–403.

Hall, Rodney Bruce and Thomas J. Biersteker. 2002. *The Emergence of Private Authority in Global Governance*. Cambridge: Cambridge University Press.

Harrison, Kathryn. 1999. "Talking with the Donkey: Cooperative Approaches to Environmental Protection." *Journal of Industrial Ecology* 2 (3): 51–72.

Haufler, Virginia. 2003. "Globalization and Industry Self-Regulation", In Miles Kahler and David A. Lake (eds), *Governance in a Global Economy: Political Authority in Transition*. Princeton, NJ: Princeton University Press.

———. 2001. *A Public Role for the Private Sector: Industry Self-Regulation in the Global Economy*. New York: Carnegie Endowment for International Peace.

———. 1999. "Self-regulation and Business Norms: Political Risk, Political Activism", In Claire A. Cutler, Virginia Haufler and Tony Porter (eds), *Private Authority and International Affairs*. Albany: State University of New York Press.

Hochstetler, Kathryn. 2003. "Fading Green? Environmental Politics in the Mercosur Free Trade Agreement." *Latin American Politics and Society* 45(4): 1–32.

———. 2002. "Assessing the 'Third Transition' in Latin American Democratization: Civil Society in Brazil and Argentina." *Comparative Politics* 35 (1): 21–42. With Elisabeth Jay Friedman.

Hoffman, Andrew J. 1997. *From Heresy to Dogma: An Institutional History of Corporate Environmentalism*. San Francisco: New Lexington.

Hovi, Jon, Detlef F. Sprinz and Arild Underdal. 2003. "The Oslo-Potsdam Solution to Measuring Regime Effectiveness: Critique, Response, and the Road Ahead". *Global Environmental Politics* 3 (3) August: 74–96.

Jack, D. 1999. *Sobre Bosques y Mercados: Certificación y Manejo Sostenible en Bolivia*. In CFV, *Boletín Informativo del CFV* 2(2). Santa Cruz, Bolivia: CFV.

Jenkins, Rhys (ed.). 2000. *Industry and the Environment in Latin America*. London: Routledge Press.

Jeppesen, Soeren and Michael W. Hansen. 2004. "Environmental Upgrading of Third World Enterprises Through Linkages to Transnational Corporations. Theoretical Perspectives and Preliminary Evidence". *Business Strategy and its Environment* 13: 261–274.

Kagan, Robert A. and David Vogel. 2002. *Dynamics of Regulatory Change: How Globalization Affects National Regulatory Policies.* Berkeley: University of California Press.

Kagan, Robert A. and Lee Axelrad (eds). 2000. *Regulatory Encounters: Multinational Corporations and American Adversarial Legalism.* Berkeley: University of California Press.

Kahler, Miles and David A. Lake (eds). 2003. *Governance in a Global Economy: Political Authority in Transition.* Princeton, NJ: Princeton University Press.

——. 1999. "Modeling Races to the Bottom." Unpublished paper, available at *http://irpshome.ucsd.edu/faculty/mkahler/RaceBott.pdf.*

Keck, Margaret E. and Kathryn Sikkink. 1999. *Activists Beyond Borders: Advocacy Networks in International Politics.* Ithaca, NY: Cornell University Press.

Keohane, Robert O. and Joseph S. Nye Jr. 2003. "Redefining Accountability for Global Governance", In Miles Kahler and David A. Lake (eds), *Governance in a Global Economy: Political Authority in Transition.* Princeton, NJ: Princeton University Press.

——, Peter M. Haas and Marc A. Levy (eds). 1993. *Institutions for the Earth: Sources of Effective International Environmental Protection.* Cambridge, MA: MIT Press.

King, Andrew and Michael Lenox. 2002. "Sustaining Industry Self-regulation in the Face of Free-riding". Unpublished Working Paper. January 2002.

——. 2000. "Industry self-regulation without sanctions: The chemical industries Responsible Care program". *Academy of Management Journal* 43 (4).

Krasner, Stephen D. (ed.). 1983. *International Regimes.* Ithaca, NY: Cornell University Press.

Levy, David L. and Peter J. Newell (eds). 2005. *The Business of Global Environmental Governance.* Cambridge, MA: MIT Press.

——. 2005. "Business and the Evolution of the Climate Regime: The Dynamics of Corporate Strategies", In Levy and Newell (eds), *The Business of Global Environmental Governance.* Cambridge, MA: MIT Press.

—— and Daniel Egan. 1998. "Capital Contests: National and Transnational Channels of Corporate Influence on Climate Change Negotiations". *Politics and Society* 26 (3): 337–361.

Lipschutz, Ronnie D. 2004. *Global Environmental Politics: Power, Perspectives, and Practice.* Washington, DC: Congressional Quarterly Press.

Lipschutz, Ronnie D. and Cathleen Fogel. 2002. "Regulation for the Rest of Us? Global Civil Society and the Privatization of Transnational Regulation", In Rodney Bruce Hall and Thomas J. Biersteker (eds), *The Emergence of Private Authority in Global Governance.* Cambridge: Cambridge University Press.

Lipschutz, Ronnie D. 1996. "Reconstructing World Politics: The Emergence of Global Civil Society", In Jeremy Larkins and Rick Fawn (eds), *International Society After the Cold War.* London: Macmillan. pp. 101–131.

Lópes, Andrés. 1997. "De la Sustitución de Importaciones al Régimen Neoliberal", In Daniel Chudnovsky and Andrés Lópes (eds), *Auge y Ocaso del Capitalismo Asistido: La industria petroquímica latinoamericana.* Buenos Aires: Alianza Editorial.

May, Peter H. 2004. Forest Certification in Brazil. Paper presented at the Symposium *Forest Certification in Developing and Transitioning Societies: Social, Economic, and*

Ecological Effects. Yale School of Forestry and Environmental Studies, New Haven, Connecticut.

McAllister, Lesley Krista. 2004. *Environmental Enforcement and the Rule of Law in Brazil*. Ph.D Thesis. Energy and Resources Group, University of California at Berkeley.

McMichael, P. 2000. *Development and Social Change: A Global Perspective*. Thousand Oaks, CA: Pine Forge Press.

Mitchell, Ronald B. 2002. "A Quantitative Approach to Evaluating International Environmental Regimes." *Global Environmental Politics* 2 (4) November.

———. 2001. "Institutional Aspects of Implementation, Compliance and Effectiveness", In U. Luterbacher and D. F. Sprinz (eds), *International Relations and Climate Change*. Cambridge, MA: MIT Press.

Murphy, David F. and John Bendell. 1997. *In the Company of Partners*. Bristol: Polity Press.

Nash, Jennifer and John Ehrenfeld. 1997. "Codes of Environmental Management Practice: Assessing their Potential as a Tool for Change." *Annual Review of Energy and the Environment* 22: 487–535.

Newell, Peter. 2005. "Citizenship, Accountability, and Community: The Limits of the CSR Agenda." *International Affairs* 81 (3): 541–557.

North, Douglas. 1990. *Institutions, Institutional Change, and Economic Performance*. Cambridge: Cambridge University Press.

O'Neill, Kate. 2004. "Privatizing the Environment: Private and Hybrid Governance Regimes at the Global Level". Paper presented at the Annual Meeting of the American Political Science Association, Chicago, Illinois, September 2–5, 2004.

O'Rourke, Dara. 2003. "Outsourcing Regulation: Analyzing Non-Governmental Systems of Labor Standards and Monitoring." *The Policy Studies Journal* 31 (1).

Perry, Martin and Sanjeev Singh. 2002. "Corporate Environmental Responsibility in Singapore and Malaysia", In Peter Utting (ed.), *The Greening of Business in Developing Countries: Rhetoric, Reality and Prospects*. London: Zed Books.

Porter, Tony. 1999. "Hegemony and the Private Governance of International Industries", In Claire A. Cutler, Virginia Haufler, and Tony Porter (eds), *Private Authority and International Affairs*. Albany: State University of New York Press.

Porter, Michael and Class van der Linde. 1995. "Green and Competitive: Ending the Stalemate." *Harvard Business Review*. September–October: 120–134.

Porter, Michael. 1990. *The Competitive Advantage of Nations*. New York: Macmillan Press.

Prakash, Aseem and Matthew Potoski. 2006. "Racing to the Bottom?: Trade, Environmental Governance, and ISO 14001". *American Journal of Political Science* 50 (2) April.

———. 2005. "Green Clubs and Voluntary Governance: ISO 14001 and Firms' Regulatory Compliance". *American Journal of Political Science* 49 (2).

Prakash, Aseem. 2000a. *Greening the Firm: The Politics of Corporate Environmentalism*. New York: Cambridge University Press.

———. 2000b. "Responsible Care: An Assessment." *Business & Society* 39 (2) June.

Pratt, Lawrence and Emily D. Fintel. 2002. "Environmental Management as an Indicator of Business Responsibility in Central America", In Peter Utting (ed.),

The Greening of Business in Developing Countries: Rhetoric, Reality and Prospects. London: Zed Books.

Quevedo, Lincoln. 2005. "Forest Certification in Bolivia", In B. Cashore, F. Gale, E. Meidinger, and D. Newsom, (eds), *Confronting Sustainability: Forest Certification in Developing and Transitioning Countries.* New Haven, CT: Yale School of Forestry and Environmental Studies Press.

Reinhardt, Forest. 2000. *Down to Earth: Applying Business Principles to Environmental Management.* Boston: Harvard Business School Press.

República Argentina, Ministerio de Desarrollo Social y Secretaria de Ambiente y Desarrollo Sustentable. 2002. *Primer Inventario Nacional de Bosques Nativos.* Buenos Aires: Secretaria de Ambiente y Desarrollo Sustentable, Dirección de Bosques.

Rodrik, Dani. 1999. *The New Global Economy and Developing Countries: Making Openness Work.* Washington, DC: Overseas Development Council, Policy Essay No. 24.

Rosenau, James N. and Ernst-Otto Czempeil (eds). 1992. *Governance Without Government: Order and Change in World Politics.* Cambridge: Cambridge University Press.

Sabel, Charles, Dara O'Rourke and Archon Fung. 2000. *Ratcheting Labor Standards: Regulation for Continuous Improvement in the Global Workplace.* Available at: *http://www2.law.columbia.edu/sabel/papers/ratchPO.html.*

Sabsay, Daniel A. 2004. Constitución y Ambiente en el Marco del Desarrollo Sustentable. Initial panel presentation at the *Symposium of Latin American Judges and Regulators: Application of and Compliance with Environmental Standards.* Buenos Aires: Fundación Argentina de Recursos Naturales (FARN).

SAGPyA (Secretaría de Agricultura, Ganadería, Pesca y Alimentos). 2004. Diagnóstico del sector forestal al 2003. Powerpoint report available at: *http://www.forestacion. gov.ar/.*

Sánchez Acosta, Martin. 2000. "Recursos Forestales Argentinas", In L. Corindalesi, L. La Rosa, S. Brandon, D. Pinasco, and C. Frisa (eds), *Argentina: Sector Forestal.* Buenos Aires: SAGPyA. December.

Sánchez, Luis Enrique. 2002. "Perspectives on Sustainable Development in Brazil." In *Encyclopedia of Life Support Systems* (copy shared with author).

Sassen, Saskia. 1996. "Losing Control? Sovereignty in an Age of Globalization." In *Columbia University Leonard Hastings Schoff Memorial Lectures.* New York: Columbia University Press.

Schorr, Martin. 2004. *Industria y nación: Poder económico, neoliberalismo y alternativas de reindustrialización en la Argentina contemporánea.* Buenos Aires: Edhasa.

Schmidheiny, Stephan and Federico Zorraquin. 1996. *Financing Change: The Financial Community, Eco-efficiency, and Sustainable Development.* Cambridge, MA: MIT Press.

Sell, Susan K. 2003. *Private Power, Public Law: The Globalization of Intellectual Property Rights.* Cambridge: Cambridge University Press.

Seroa da Motta, Ronaldo. 2002. Determinants of Environmental Performance in the Brazilian Industrial Sector. Working Paper in series: *Regional Dialogue on Policy Issues.* Washington, DC: Inter-American Development Bank.

Simmons, Beth and Lisa Martin. 1998. "Theories and Empirical Studies of International Institutions." *International Organization* 52 (4) Autumn : 729–757.

Smart, Bruce (ed.) 1992. *Beyond Compliance: A New Industry View of the Environment.* Washington, D.C.: World Resources Institute.

Smith, D. A., D. J. Solinger and S. C. Topik. 1999. *States and Sovereignty in the Global Economy.* London: Routledge Press.

Snidal, Duncan. 1979. "Public Goods, Property Rights, and Political Organization". *International Studies Quarterly* 23 (December): 532–566.

Sprinz, Detlef F. and Carsten Helm. 1999. "The Effect of Global Environmental Regimes: A Measurement Concept". *International Political Science Review* 20 (4): 359–369.

STCP. 2000. *Plan Estratégico para el Desarrollo del Sector de Bolivia.* Curitiba, Brasil. Available at: *http://www.cadefor.org/en/sectfor/eplan_en.*

Stein, Arthur. 1982. "Coordination and Collaboration: Regimes in an Anarchic World." *International Organization,* 36 (2).

Steven Bernstein and Benjamin Cashore. 2008 "The Two-level Logic of Non-State Market Driven Governance". In Volker Rittberger, Martin Nettesheim and Carmen Huckel (eds), *Changing Patterns of Authority in the Global Political Economy: Volume II: New Actors and Forms of Regulation.* Palgrave Macmillan.

———. 2007. "Can Non-State Global Governance be Legitimate? A Theoretical Framework". *Regulation and Gover nance* 1: pp.1–25

Strange, Susan. 1996. *The Retreat of the State: The Diffusion of Power in the World Economy.* Cambridge: Cambridge University Press.

Swire, Peter P. 1996. "The Race to Laxity and the Race to Undesireability: Explaining Failures in Competition among Jurisdictions in Environmental Law." *Yale Law and Policy Review* 14: 67–110.

Tiebout, Charles M. 1956. "A Pure Theory of Local Expenditures." *Journal of Political Economy* 64 (5) October: 416–424.

Utting, Peter (ed.). 2002. *The Greening of Business in Developing Countries: Rhetoric, Reality and Prospects.* London: Zed Books.

Vogel, David. 2006. "The Role of Civil Regulation in Global Economic Governance." Paper presented for the *Global Economic Governance Programme.* Oxford University.

———. 2005. *The Market for Virtue: The Potential and Limits of Corporate Social Responsibility.* Washington, DC: The Brookings Institution.

———. 1995. *Trading Up: Consumer and Environmental Regulation in a Global Economy.* Cambridge, MA: Harvard University Press.

Weidner, Helmut and Martin Jänicke (eds). 2002. *Capacity Building in National Environmental Policy: A Comparative Study of 17 Countries.* Berlin: Springer.

Wongtschowski, Pedro. 2002. *Indústria Química - Riscos e opportunidades.* 2nd Edition. Editora Edgard Blücher.

Young, Oran R. 2001. "Inferences and Indices: Evaluating the Effectiveness of International Environmental Regimes". *Global Environmental Politics* 1(1) February:. 99–121.

Young, Oran R. and M. A. Levy (eds). 1999. *The Effectiveness of International Environmental Regimes: The Causal Connections and Behavioral Mechanisms.* Cambridge, MA: MIT Press.

INDEX